The Biblical Flood

The Biblical Flood:
Global Warming & Bush's Harvest

William W Morgan

BookSurge Publishing,
an Amazon.com Company
Universal Salvage Corporation
Specialists in Dying Civilizations
http://universal-salvage.org,uk

Published by
BookSurge Publishing,
an Amazon.com Company
Universal Salvage Corporation
Specialists in Dying Civilizations
239 S. Juniper Street
Escondido, CA 92025
U. S. A.
SAN: 850-4016

ISBN 0-9778492-3-6 (paperbound)
LCCN: 2006908902
First Edition: 2006

Printed in the United States of America

Contents

About the Cover i
Gratitude iii
Preface v
Chapter 1: Background 1
Chapter 2: Natural and Manmade Disasters 11
Chapter 3: Ascent of Man 23
Chapter 4: Descent of Man 47
Chapter 5: The Natural Society 73
Chapter 6: Natural & Societal Disasters 119
Chapter 7: Solution to Our Problems 145

Appendices

A Definitions 183
B Quotations 195
C Bibliography 205
D Achievements 211

Author of:
Cowardly New World
ISBN: 1-4196-2610-8
<http://universal-salvage.org.uk/book_list.php#cnw>

An "Einstein" Thought Experiment
ISBN: 0-9778492-0-1
<http://universal-salvage.org.uk/book_list.php#ein>

On… Terrorism: Simple Cause — Simple Cure
ISBN: 0-9778492-4-4
<http://universal-salvage.org.uk/book_list.php#on>

About the Cover

The cover is my idea. Including much of the details. So if you want to blame someone—to accuse someone of heresy, blasphemy, or being polemic—you know who to accuse and charge.

As usual my friend Billy Tudhope implemented my ideas. You may compliment him on his artistic and graphical expertise or not as you wish. But please do not blame him for the content and the statement it makes. This is totally for my account.

He does not agree with everything I think and believe so if there is something you do not like I am the one to blame or accuse and not him.

I think that certainly within one hundred years the sea level will be at least as high as indicated. And perhaps much less time and much higher. And the sea level rise, associated climate changes and societal chaos that will occur is totally for the account of people who occupy such buildings. They are the cause of all our societal problems. Including the chaos that will come from the effects of Global Warming. They are the cause and cannot be part of the solution.

Not being able to make anything work in society, how could they possibly handle the problems of Global Warming?

Gratitude

First and primarily gratitude must be given to my great and wonderful teacher Andrew J. Galambos (1924-1997). Who, among other things, taught me how to think. To understand the concept of positive history and "how to know you are right". To evaluate knowledge for rightness.

Galambos also introduced me to many other teachers. Those in the flow stream of history who have created new and useful knowledge and possibly tangible products. Those from Archimedes to Zemansky.[1] From the time of about 500 bce to yesterday and even today.

Next, and also at the top of the list, my great and wonderful wife. Who has intelligence, understanding and patience beyond measure. Putting up with an angry, cantankerous old man. She is a person of unique, indispensable qualities. My best friend in life.

And for this book, the individuals I will label as "climatologists". Although this covers individuals in many different technical disciplines.

I hesitate to list individuals. Simply because there are some whose identity I do not even know. And many I do but may not have their name at hand. Yet I feel I should list those that I do know, and will probably do so in an appendix to this book.

As often happens, there are some who did work decades ago, the value of which is only now being understood. And some who made a major contribution only yesterday. By brilliantly "tying all the loose ends together". And probably starting a very valuable new stream of thought for thinkers who will suddenly become very important tomorrow.

As I will probably say in that planned appendix, you should enter their name into a search engine, such as Google. Just their name (in quotes) will produce some references.

---

1 Mark W Zemansky, one of the great thermodynamic scientists.

Then you might want to narrow the search based on operative terms of interest.

Preface

Many people will say this is too "polemic". That it is not scholarly. The latter is certainly true. That I have obscured or destroyed the possibility of transmitting an important message by the way I have written this.

But I think I almost want you to become angry at me. So angry you will read with the intent of trying to understand what I say so that you can refute it. I want you to become angry and to *think*.

It is very, very important that we understand what sea levels and temperatures might reach at a certain time. What each of these levels might do to the fabric of society. What the worst could do and how soon that might happen.

And to then toss out the information. And assume that we must cut back our particulate and greenhouse gas emissions to a certain level within five years. Maybe one that might return us to the 1950 level?

Why?

Because the cost of being wrong and not doing this *will be specie extinction.* We will almost certainly arrive at a time when we suddenly realize that "things are out of control". That there is *nothing* we can do to prevent our extinction. We can wish we had done something but we can't and it is too late. And we will know with absolute certainty that each of us, individually, will physiologically suffer an agonizing death.

And there will be those of us who will also suffer in another way. Knowing that as a life form we spent 3.65 billion years to become a thinking specie. And about 7 million years as the genus Homo.

We had discovered and articulated many if not most of the laws of nature. The ability to create wondrous technological devices to make our life continuously better.

But we created a societal system that was a disaster. Not allowing us to solve even one of our basic societal problems. Rather, they became ever-more endemic and intractable.

A societal system that will actively mitigate against the solution to the problems of Global Warming. It will be this system that—unless we get rid of it and adopt a rational and durable societal system—is the real cause of our extinction. And not Global Warming. We could solve those problems.

Where will you go if you live in Amsterdam, Bremen, Boston, Hamburg, London, Los Angeles, Miami, Osaka, San Diego,, Washington DC? The water is not yet lapping at your feet. But you know it is coming. And so do others. As property values are plunging. Trade is at a standstill and there is talk of further unemployment.

The sea level has risen a half meter—a foot and half—and it is certain it will rise three meters or about ten feet. But there is talk of as much as sixty meters or almost two hundred feet. As the Greenland Ice Cap is melting. Possibly even the ice at the Poles. No one knows for sure yet, but the signs are not good.

Including the ice storms that have been plaguing most of North America and Europe. And there is talk that this may switch and the areas—including even Central America—will become virtual deserts.

Do you have a passport? Will your proposed destination give you a visa? If you haven't already sold your house and/or business now it is unlikely anyone will buy it. Or the speculators that might will not give you very much for it.

They talk of being able to first slow and then halt the changes. And some visionary scientists—now free of the control of political overlords—are talking of even being able to reverse some of the worst effects. Although not before there is extensive damage to the infrastructure and the fabric of society.

The disorder in society is threatening to tear everything apart. Including the work of the scientists who might fix things.

I know that I will be roundly criticized for painting this Apocalyptic picture. But I think it best to tell the truth. About what will probably happen. Whether it is in five or ten years or in a hundred years.

And this is not the worst news. The really bad news is the societal disorder and technical problems related to Global Warming and climate change can not be changed with our present societal structure. A societal structure based on coercion.

What do you see when you read a newspaper or news magazine or watch the daily news on TV. What do you see? Societal problems that are becoming ever-more endemic and intractable. This happens because our societal system seeks to violate the laws of nature. Those laws of nature that govern and explain the societal behavior of thinking beings.

It has been going on for at least 5,000 years. And has been getting worse the entire time. Read and think about the history of our specie as a societal animal. We have solved countless technological problems. But not one societal problem.

What this tells us is there is no hope of solving the problems related to Global Warming unless we change our societal system. If it continues to be based on coercion, we will perish as a specie. Sooner rather than later with added problems of the societal disorder that will result from Global Warming.

We probably have only a couple of years left to start changing.

In a recent article on the status of Global Warming I read something that struck an alarm bell. The ignorant, fallacious and self-serving statement that something so important as the handling of the problems related to Global Warming can not be left to the "private sector". Meaning it must be done by politicians.

Remember, these are the people who have brought you:

System or Service	Result
School system	35% of the school leavers are functional illiterates. System costs more every year, produces lower quality.
Pension system	Financially unsound with ever-increasing unfunded liabilities. A "Ponzi" racket, where money to support operations comes from an ever diminishing number of future customers.
Healthcare system	Amidst great advances in biological sciences and technology, healthcare becomes worse and more expensive.
Personal Security	The régime is supposed to provide you with personal security. Do they succeed?
Inflation	Only a money issuer can create inflation. By debasing the value of the currency. Price increases necessary as a result of this are just that: price increases. And *not* inflation.

The government was supposed to protect the people and city of New Orleans from natural disasters. This was a relatively small scale event compared to those that will come in the future.

If you read the US government plan to protect citizens and property, they propose to solve the problem by:

- Raising the beaches.
- With levees and dikes.
- And pumps.

This is what was supposed to protect New Orleans from Katrina. And the organizations they used were FEMA and the Corps of Engineers. They plan to use the same "organizations" (they are organized…?) to protect against all aspects of Global Warming. Do I need say more…?

In January 2005 there was heavy flooding in Carlisle, England. Perhaps up to 10,000 homes involved. More than a year and half later at least 3,000 people have still not been able to return to their homes. And there is talk of a 10-15 *year* rebuilding program. Although for the people of Carlisle this is their universe, in planetary terms this is very insignificant. Compared to what will happen with the sea level rise that will occur within thirty years or so. Effecting hundreds of millions of people.

Under the societal system based on politics we can not solve even one of our societal problems. Even ones that in a sense only cover a small part of the activities of thinking beings in general. Not to speak of something involving the entire population of this planet.

What do you think the result will be of something that involves all aspects of human endeavor? What do you think the chances are that Blair, Bush, Chirac and their kind can solve the problems resulting from Global Warming? And this is not like the matter of the post office, train system or healthcare system not working. Somehow—in spite of, and not because of politics—we can muddle though. And send the letter again or call the person. Or somehow get your medical problem handled or chose an alternate transit system. Or not go at all.

This time we are talking about a "one chance" shot at handling Global Warming. And if we do not do it successfully, *we will perish as a specie.* Sooner rather than later. And in great pain and discomfort. Besides the physiological pain, the mental pain of knowing "we blew it", as the Yanks say. We had a chance to fix things. Did not, and now we will suffer the ultimate disaster for any living, thinking being. To die. There will be no next generation to carry on. A million years from now extraterrestrial archeologists may have a hard time establishing if there was ever life here beyond bacteria, viruses and insects.

The one thing the politicians should be good at—war—they can't get right either. With thousand of years of history on which to base their actions.

Politicians have a position of perceived or pseudo "importance". One that lasts only as long as they occupy the position of power to which they have descended.[2] And ends once they have stepped up from office.

"Importance" is in quotes because they have no real importance. As importance is a measure of the amount of property morally created. Both tangible and intangible. Useful, tangible products or services, and new and useful knowledge. Their position is one more of notoriety rather than one of importance.

Ask Jimmy Carter, Bill Clinton or even Tony Blair about this. Blair knows he is "stepping off of the world stage" and into a position of clear and obvious irrelevance. Politicians are only relevant when they can control the lives and property of others. With their relevance being that they degrade the lives of all their victims. And in these days of "globalization" and "superpowers" this can include anyone on the planet.

If you could ask the tragic Ariel Sharon he would tell you the same thing. One day virtually the entire world hung on his every word and deed. Then he had a stroke. And within one day he became irrelevant in the world.

Think of Thaksin Shinawatra, the former de facto ruler of Thailand. When he arrived at a recent UN meeting he appeared to have significance in the world. The next day, if he called George Bush the telephone operator would have said "… what is your name again"? "Oh, I'm sorry, Mr. Bush is busy at the moment". In hours he became irrelevant. Why? Because he never had *true* relevance.[3]

Contrast this with someone like Jonas Salk, who created a successful polio vaccine. Or William Hewlett and David

---

2 Descend is the correct direction. As they are going from a situation in which they probably could not control the lives and property of others to one in which they must and will do so. With all of their actions being immoral and degrading the quality of life of their victims. When a politician leaves office they "step up", rather than down.

3 In some ways this is a poor example. Because he is a successful business man, and in that world he has true relevance and importance. It is a bit like Ronald Regan. He had importance as a film actor. But he threw that all away when he descended to political office.

Packard who, from about $600 or so, created a company that employs millions of people. With medical products that have saved and enriched the lives of very many people. Or the unknown Chester Carlson, who in his garage by himself created the first successful photocopy machine. And a product and industry that totally changed the world for the better. Even when long dead these people still have relevance and importance.

Chester Carlson's life overlapped that of Lyndon B. Johnson. Which one do you think was important to Homo sapiens? And which either irrelevant or the world would have been better without them? Can we generalize this for all the politicians who ever lived?

Contrast these people and their accomplishments with the pseudo-accomplishments of the politicians as listed in the earlier table. And remember that decades after the accomplishments of Salk, Hewlett, Packard and Carlson we still enjoy the result of their lives. Even when they are long dead. While we live in misery under the societal system of politics and politicians. And often suffer from it for years after they are gone. That is the legacy of them and their failed societal system.

Assume there is "reform" every 4-5 years. The usual "term of office" of a politician. This means in the past 5,000 years there have been a minimum of 1,000 reforms. But then you must remember to multiply that by the number of political entities and "ruling politicians". And when you consider all the towns and villages in the world that is millions. Hundreds of millions of attempts to make the system work. And as we know, nothing is working now, yes?

Maybe it is time we thought about getting rid of the system? Millennia of history shows that changes to it have never worked. Proving it is an unworkable concept.

Wouldn't you say that maybe Global Warming is too important to be left to the politicians? To put it in their hands? Or for them to have anything to do with it? Might it not be better for each individual to begin assuming responsibility for their own life? Assume the sole responsibility for the duties and obligations of individual thinking life?

At the very least the individual can then say: "I did my part. I acted morally and rationally. And on average I was at or below my allowable emission limits". And if you can say that soon, then the worst will *not* happen. And we will become a rational and durable specie. With our time scale of existence being concomitant with that of our own local universe. Whether we chose to remain here on the planet Aristarchus or to disperse throughout the universe.

Please be aware and understand that I will sometimes repeat principles and situations many times. It may be irritating to you. And you may tend to skip over them. But please do not. I do it because many concepts are new, unfamiliar—even unpleasant. But they are reality. And repeating and saying the same thing in a different way or under a different situation may help to overcome a deeply ingrained fallacy.

Chapter 1

Background

Synopsis

We now know that Global Warming is *real* and that it is accelerating. That it will cause rising sea levels. That temperatures will increase. That desertification will increase. We do not know the amount of these things and exactly when some problems will appear and where. Only that it will not be good for Homo sapiens. The probability of societal extinction is real, and higher than the probability of survival. For reasons that should become apparent as you read this book.

The general facts related to Global Warming are now agreed by all mainstream scientists. Those who do not agree are in denial of the facts and further threaten our very survival as a specie.

We can only create survival for Homo sapiens by properly handling the technical and societal problems. If we do not change our societal structure from one based on coercion to one based on individual freedom we will *not* be able to handle the technical or societal problems related to Global Warming. And *will* perish as a specie.

Background

With new information now becoming available on Global Warming and its potential consequences I have become almost too depressed to finish this book. Yet I know I must, as the information provides the knowledge necessary to work our way out of the problem. And it is the *only* solution that will work. With "the problem" being the extinction of the specie Homo sapiens. That's us.

We face two problems, and Global Warming is actually the least worrisome of the two. The real problem is Societal

Ignorance, or "Global Dumbing". As represented by the fact we have never solved even one of our basic societal problems. That they become ever-more endemic and intractable. That no one seems to know the cause of our societal problems and thus can not solve them. Well, no one except me. Which is why this book on Global Warming is so important. Because it describes the cause and cure for our societal problems.

We can not solve them with our current societal system. And that if we do not first solve our problem of Global Dumbing we will *not* be able to solve the societal problems that will result from the effects of Global Warming. *And in the absence of any Global Warming problems we would still become extinct.* If we do not fix out societal problems.

And the 3.65 billion year journey that led to Homo sapiens will have been for naught. With us being the cause of that. Our own 7 million year journey as hominids ending in disaster.

And disaster not just for us. The coup de grace for us will be the release of methane gas from the currently-frozen methane hydrates at the bottom of the oceans. When conditions are "right" to start that *there will be no stopping it.* Temperatures will reach a level not seen on the planet for 4 billon years. And probably all life except for some microbes will vanish from this planet. With thinking life never to occur again on this planet. As it will be too late. Another period of 3.65 billion years—assuming it took about that long—would bring the time very near the time for Copernicus—our local star—to start dying.[4]

It was only on August 27th 1859 that oil began pouring from what had been termed "Drake's Folly". Edwin L. Drake drilled the first oil well in the US. Starting the US oil industry.

---

4 The name given to our local star by Andrew J. Galambos (1924-1997). In honor of the creator of the heliocentric hypothesis. While wrong, it did remove Aristarchus, our planet, from the center of the universe. And was thus a vast improvement over earlier hypotheses.

Everyone had said it was "folly" to think that oil could be obtained from the ground. It turns out they were right, but in an unexpected way

And ten years later in 1869 when Thomas F. Rowland patented the concept of the offshore oil platform. Although our societal system based on coercion in itself dooms us to early extinction, ignoring the consequences of burning fossil fuels will greatly accelerate the process. But it may be a close race. Which problem would lead to our extinction first: our faulty societal system or the societal chaos that will result from Global Warming.

It is less than 150 years from that fateful day. Yet the motorcar would not appear for another fifty years. And industry—with it the big electric generating stations—probably did not really get started until a hundred years ago. Ironic that the creation of the "clean energy" of electricity required the use of coal and the release of vast amounts of particulate matter and CO^2 gas.

Correct me if I am wrong, but it was probably only with Rachel Carson's publication of *Silent Spring* in 1962 that we began to look at what we were doing to our environment and ourselves?[5]

So we have actually discovered the disease fairly quickly. The problem is that it was both a "silent killer" and a very virile disease. And the patient—Homo sapiens—is now in *very* critical condition. In a historical sense seconds away from death. And if we do not marshal *all* of our defenses it *will* occur. By "defenses" I mean current useful knowledge, and new and useful knowledge that we can and will create to solve the problems.

We must do whatever is necessary to keep the temperature of the patient somewhat below the temperature

---

5 You may send me a message using the following link: <http://universal.salvate.org.uk/ReaderMessage.php>.Hackers, please it is not in your own best long-term interest to abuse this knowledge. Turn your abilities to our survival as a specie. Which includes you.

at which the methane hydrates would start releasing methane. If that should start there is no solution. And within a few years we—you, I and the specie—will die a painful death.

"Whatever is necessary means" *heroic measures*, but not immoral measures. Because immoral measures—the current concept of a societal system based on coercion, with its unnatural laws and political police to enforce them—is what has gotten us into our current situation. The concept of "politics" and of "political action". One or a few people controlling the lives and property of all others.

This is the societal structure that has not allowed us to solve even *one* of our basic societal problems. And which will actively mitigate against solving the problem of Global Warming. With its "hidden companion", Global Dimming.

Politics and political action can not make post offices, pension systems, healthcare systems and schools work. They can not solve the problems of crime, terrorism, hunger, famine and malnutrition. They can not make British Rail, the National Health Service (NHS), London Underground or Amtrak work.

We argue about stem cell research, euthanasia and the right to die, about whether two people of the same sex can be married. About how long to let the Israelis murder people in Lebanon and Palestine. Confirm the Israeli "right" to continue operating the vast concentration camp known as Palestine.

Afghan, African, American, Iraqi, Christian, Shiite, Somali, Sunni, and other warlords run wild. The Americans like to call their fighters "soldiers," while those opposing them are fighters, insurgents, rebels and other such things. The American régime is referred to as an "Administration" because it is "democratic". The American ruler lives in a white-colored house, while those who oppose him live in a "presidential palace".

It might be useful to think of the truth and reality of the "American Revolution". That George Washington and others were opposing the "legally constituted" government of the time. That they were thus rebels and traitors. OK, yes they were fighting for their freedom. But are not the Afghans and Iraqis doing the same thing? And was not the régime of Saddam Hussein the "legally constituted" government of Iraq? He became the ruler of Iraq by the system in use for at least 5,000 years.

The same way George Washington ended up as president of the US. The insurgents—his soldiers—defeating those of the legally constituted regime. Thereby allowing a new regime to appear.

Politics has three basic forms: a democracy, a dictatorship or a monarchy. But in practice and principle these are only a style on the concept of politics. Controlling their victims—usually referred to as citizen—by a system of unnatural laws enforced by a political police. Well, they do work under the system of politics and are thus political police, yes?

How many years and how many millions have died during the fifty years in which politics could not end the "Palestinian problem"? Get them out of the world's largest concentration camp. It is this failure that has given birth to modern terrorism. Both of bin Laden and of Bush. Punish the criminals who caused this. The solution is actually very simple. Yet in fifty years nothing has happened. We will not have anything approaching fifty years to fix the Global Warming problem.

If the American colonists had lost the war and Washington, Paine and others had been caught they would certainly have been tried as traitors and executed. But although they were rebels they did not lose. And they thus replaced the British as the legally constituted régime. And as they say, "the rest is history".

While the mind and word games go on, millions starve in Africa. Largely as a result of the pollution from the USA that—beginning in the 1980s—meant the monsoons in the Sahel did not occur. The rain clouds did not move north from the equator. With decades of drought that killed millions. But this was okay, as for the American ruler to do something about the pollution would have been "bad for business". And for his continued employment. The "perks" of Air Force One, foreign travel without the traffic jams and airport queues. Housing and meals provided in a white-colored house that many would consider a palace.

Our faulty societal system allowing him to use his private militia to attack the country of a ruler that may have tried to harm his daddy. With the result that more than 100,000 people have been murdered. But this seems acceptable to society as he would also bring the hallowed concept of democracy to his victims. Bring them the American version of "freedom" whether they want it or not. The natural condition of thinking beings is to be free. You can only take it from them, not give it to them.

The "collateral damage" of the millions of liters of fossil fuels needed to fuel the tanks and warplanes was also acceptable, yes? Along with the billons of tons of particulate matter and greenhouse gases that were also created.

Other items in the news: trade talks have ended and trade wars continue. Alternately encouraging and prohibiting exports and imports based on every criterion imaginable. And some that no one could imagine. Subsidies are criticized and eulogized. No one can agree whether 50,000 or 100,000 sneakers can be imported. Or no sneakers. Or change the name and start the negotiations over again. Orwell's concept of *doublethink* reigns supreme.[6]

The world's "greatest"—and presumably most wonderful—democracy also has the world's largest per-capita prison population. And the political ruler of this group finds it necessary to "circumvent" the law and send people he does not like out of the country for torture. One who "circumvents" laws is usually referred to as a criminal.

And if they were not being tortured, why was it necessary

---

6 George Orwell (1903-1950). Who created the concept of *doublethink*, saying that "Doublethink means the power of holding two contradictory beliefs in one's mind simultaneously, and accepting both of them". He also said "In our time, political speech and writing are largely the defence of the indefensible".

to send them outside the country? If it was simply a security matter, are the great "Supermax prisons" not secure? And with comments like the preceding about someone who muzzles climatologists for not making him look good, I think my own freedom and security might come into question.

As far as I know, no native-born American has yet been sent to Guantanamo. But the régime has shown that it does not adhere to the law and follow precedents if it is inconvenient.

This would seem to be a diversion from the problems that will result from climate change. But it is not a diversion. It is necessary to understand that our current societal system does not work and why it does not work. And why we can not use it or allow it to be present as part of the solution to Global Warming.

Politics is a problem creating system. And not a problem solving system. Again you need only look at *results*. The world around you. More and more broken news. As the politicians flit from one disaster to another. Announce a solution within ten years and then abandon that disaster for the next one. To be discussed over sumptuous "working dinners" with the best wines. And with security costs alone vastly exceeding that needed to start any one of the major corporations around the world.

And it is these often-hated corporations and business people who bring you the good things in life. Cell phones, movies, iPod music players, computers, central heating and air conditioning, air travel. Which unfortunately adds to pollution. But technically we could do something about that.

Remember also that it is the political state that says they will regulate the people who provide you with these things. To make sure that you get what is good for you. And they fail in this also, yes?

I am not a climatologist. I will try to provide you with some information—more a summary, really—related to Global Warming and its possible effects. Although the people who created this knowledge are the ones deserving of the credit. And are the best to provide the knowledge.

I have included a rudimentary "bibliography" and "credit list" as an appendix. I encourage you to consult the original documents. Buy or borrow the books and read them. The same with DVDs and multimedia documents. And Al Gore's film *An Inconvenient Truth.* The title is correct and appropriately descriptive. It presents the truth, and the result of what it describes—Global Warming—will be very inconvenient. For everyone for different reasons and in different ways.

These people—the scientists I will mention and which may be listed in the appendix—are truly leaders. A politician can never be a leader. Well, only rarely. Such as a Hitler or Stalin. The others are simply followers. In our perverted political society they find what people want and promise to provide it to them. Usually "free". These are not leaders.

Leader

"Any individual who morally creates new and useful knowledge or products that have the potential for increasing the quality of life for other individuals."

The use of absolute definitions and semantic precision is very important. Without it there is no possibility of effective communication. No possibility of creating new and useful knowledge. Lack of absolute definitions and semantic precision has prevented us from solving any of our societal problems. It is why "social science" is neither social nor science. And is not a profession, but a mystic religion. This is why I will include a list of definitions in this book. So that when I use one of the operative words of the subject at hand you will know how I am using that word.

You must realize that if you placed ten people in a room and asked them for a definition of freedom, liberty, capitalism, justice, democracy, ethical and similar things, you would receive ten different definitions. And with further thought, these would be revised and you would have more than ten different definitions.

As I said, it is not possible to have effective communication with such a situation. To determine if existing knowledge is correct. To create new and useful knowledge. To solve problems.

In technology, the operative words of the supporting areas of "science" mean the same thing to everyone. Regardless of tribal affiliation (nationality), culture, religion, skin color and other such things. This is why we continuously create new and useful technologies and associated products. Why we solve problems in technology.

And it is the lack of absolute definitions and the associated epistemological processes that prevents us from solving any of our societal problems. And will prevent us from solving the societal problems that will result from Global Warming.

There is only one thing that I seem to know better than anyone else on this planet. That is the cause of our societal problems and the cure. If there is anyone else who knows this I would dearly love to meet them. Obviously there is no one currently in the news that understands the solution.

I have encountered a number of people who *think* they know, but very quickly prove they do not.

One sure sign is that their purported solution requires "politics" for its implementation. Another sign is that they do not use absolute definitions and semantic precision. And probably use "situation ethics".

Or they propose a "solution" that involves "limited government". The basic concept of politics does not allow this. It requires ever-increasing laws. What I have referred to since I conceived the idea in 1956 as *Government Without End*. This will be discussed further in the following chapters. There is an eponymous chapter on this in one of my earlier books, *Cowardly New World*. And a further discussion in the chapter *Democracy and Freedom: Mutually Exclusive Concepts* in the same book.

You may read the three-part essay on democracy in HTML format at:
<http://www.aristarchus.org/DemocracyFreedom05a.php>.

Or you can download and print the PDF version at:

<http://www.aristarchus.org/pdf/DemocracyAndFreedom5.pdf>.

To learn why our chosen societal system has never worked and can never work, please read the following chapters.

Chapter 2

The Problems

Synopsis

At present our planet is beautiful, blue and tranquil. It was not always this way and will almost certainly not be that way in the long term. And our dying star will one day finish it for good.

But now we face two manmade disasters, both of which can make us extinct in a very short time: *Global Warming* and *Global Dumbing.* Almost everyone recognizes the former but not the latter. Which will destroy us in the short term if not fixed. It greatly increases the urgency and criticality of Global Warming. We must solve Global Dumbing *before* we can hope to stop Global Warming in time.

Global Dumbing is the cause of the fact we have never solved even one of our societal problems. Why they become ever-more endemic and intractable.

Background

For a society of thinking beings there can be many different disasters. Basically they would fall into two categories. Natural Disasters, disasters that they do not themselves cause. And Societal Disasters, which are caused by conscious actions of individual members of the society or specie.

Any thinking specie—assuming they are still thinking—must prepare themselves for both types of disasters. Ideally, to prevent them from happening. If not, then to solve them before they become serious. Or develop the technology to leave a dying planet.

In this book I will write about the following disasters:

Natural Disasters

- Asteroid and Comet Impacts.
- Magnetic Field Problems.
- Storms.
- Supervolcanoes.

Manmade Disasters

- Societal Disasters.
  - Global Dumbing & Societal Fallacies.
  - Short-term thinking.
  - Lack of understanding of rightness.
  - Lack of understanding of the laws of nature.
  - Failure to fulfill our destiny of becoming a thinking specie.
- Global Warming

For the Natural Disasters I will really only mention them very briefly. As the main topic is Global Warming. Also, if we can truly solve all aspects of the Global Warming problem then we are well on the road to solving the problems related to Natural Disasters.

Yet if the main topic is Global Warming, why did I mentioned it last? After all this is the subject of this book. Although it may very well result in the extinction of our specie—and most others except insects, bacteria and viruses—it will not really be what killed us. Which is the same thing that killed all previous civilizations: coercion, or lack of individual freedom.

But this time, because of "the globalization of politics" it is not a single political entity that will fail. It will be the civilization of Homo sapiens.

I suppose I should not mention now what should probably be a conclusion. It seems almost too simple, It is simple, Yet not simplistic. Nature and our universe is basically simple. We comprehend it only slowly, which is what makes it seem non-simple. No more on that now, but please do think about it…?

What would doom us to extinction is the last item under

societal disasters: our failure to think. As exemplified by the four preceding items. How do I know this is true? How can you verify that it is true? By taking a cursory glance at history and listening to the daily news. And then *think*. Think of what has been the result of our chosen societal system. One based on coercion. Whether of a democratic, dictatorial or monarchial style.

We have never solved even one of our societal problems. Not one. Rather, they become ever-more endemic and intractable. Almost reaching the point where the time devoted to "broken news" exceeds that for "unbroken news". And a thinking being must wonder if this will be the destiny of our specie. If our epitaph will be "They died in and of ignorance".

How do you expect us to solve the societal chaos that will result from the effects of Global Warming? If we can't handle a relatively very small thing such as Katrina? And that will be small relative to the things in store for us. In January 2005 there were floods in Carlisle, England. Now, more than eighteen months later, a large percentage of the families are still waiting to get back into their homes. And by future standards this was a very, very minor event. There were only about 3,000 properties effected.

What will happen when it is in the hundreds of thousands of properties…?

Natural Disasters

Asteroid and Comet Impacts

As with the other natural disasters we know these events—asteroid and comet impacts—will happen but not when. And we can probably only make a guess at their severity. Although as population density has increased since the last major impact and is increasing all the time, even a "tiny" meteorite could cause a huge loss in lives and property.

We also know—or should know—that as our knowledge increases we can sooner and better detect comets and asteroids that might impact Aristarchus.[7] And also be able to alter their trajectory so they do not.

So if we have a few more years—maybe a couple of decades at most—before such a pending impact we can probably get rid of this problem.

Supervolcanoes

These have come to our attention only recently. And we now know that a volcanic eruption about 75,000 years ago very nearly wiped out our specie. It has been estimated that at most about 10,000 Homo sapiens may have survived. And maybe as few as 2,000 or so.

There are at least a couple of Supervolcanoes around. The aforementioned one was what created Lake Toba in the middle of the northern part of the Indonesian island of Sumatra. And that while there is no apparent volcano there now the magma is lurking below the surface. And will almost certainly erupt again some time in the future.

Another Supervolcano seethes below Yellowstone National Park in the western United States. The caldera lurking under the serene park is about 20 x 50 kilometers in size. In other words, about 1,000 square kilometers in area. For Americans, 12 x 30 miles or about 360 square miles in area.

It last erupted about 640,000 years ago. And on average has erupted about every 600,000 years. So on *average* we are overdue by about 40,000 years. The 600,000 year period roughly corresponds to the time our specie has existed. We did not exist as a specie at the time of the last eruption.

But we have an idea what would have happened had we existed. Only the "local effects" indicate that animals thousands of miles away would be inundated by choking layers of ash. Most of what is now the United States was buried under

---

7 Again, this is the proper name for our planet, so-named in honor of Aristarchus of Samos (310-250 bce). Geometer, astronomer and mathematician, he was the first to reasonably accurately determine the size of the earth and moon and distance to Copernicus, our local star. He also knew the planet revolved around Copernicus.

several *feet* of ash. The rest of the world would fare better, but not well.

Within hundreds of miles your next breath would be your last. Your lungs lacerated by minute, glass-like particles. Perhaps within a thousand miles or more. Imagine a few feet of snow falling within a short period of time. With snow you could open the door and have fresh, cool air. With this you would not have air you could breath for days. Perhaps weeks or years.

Because of the population density of the planet, the need for food and other such things I think it unlikely that other than a very hardy band of people perhaps living underground would remain. Or maybe some of the Indians of the Andes might survive.

The effects would be so devastating though and we have so little we could do to prevent it that it is hardly worth thinking about?

Yet this says there is a limit to what Homo sapiens can learn and do. If a hundred years ago we knew with certainly that we would soon be hit by a large asteroid or comet we would have said there was nothing we could ever do to handle this problem.

I think it almost certain that given enough time before such an event—say maybe a hundred years—that we might learn how to relieve the magma pressure that causes an eruption. I do not propose to speculate how, since I have no direct, detailed knowledge in the relevant scientific disciplines.

Having said that, a thought occurred to me. Maybe an "army" of nanotechnology robotic "drills". Acting as one or more very large drills. Operating autonomously on "autopilot." Able to drill many kilometers into Aristarchus. Creating a large "bore hole" allowing a "controlled breakthrough" of magma. Rather than a quick, catastrophic explosion or eruption. Perhaps through several bore holes at once. And relieve the pressure?

We might even be able to drill in from the ocean and thus have a handy "fire extinguisher". To eliminate or blunt the atmospheric effects. Maybe even creating an island that years in the future would be useful. Or even channel it back into the planet's core? Yet the technological challenge for controlling such a "safety valve" would be daunting to say the least. Fraught with the possibility of creating the very thing you were trying to prevent.

Yes, billions of nanotechnology robots would have "sacrificed their lives for us." As they would be cremated during the controlled eruption. Fanciful, I know. But sometimes the outrageous becomes pedestrian technology.

When done a thousand years in the future people would say "Well, yes, obvious. How else would you do it?"

Magnetic Field Problems

It is known that the magnetic poles of the planet have switched. And for periods of time there has been little or no magnetic field. This latter is much more disastrous than which is "north" and which is "south".

Even the origin, source or cause of Aristarchus' magnetic field is not known for certain. There are several hypotheses on the subject. But as yet no theory, else by definition we would know the origin, source and cause. And as with Supervolcanoes, perhaps even how to control this phenomenon.

As noted earlier, the absence of a magnetic field during its change or for some other reason could be catastrophic. For it is Aristarchus' Magnetosphere that protects it and us from the harmful effects of solar radiation.

And it is also this Magnetosphere through which nature puts on a beautiful light show for us: The Auroras. The "Northern Lights".

Not only could the increased solar radiation give us a severe sunburn (or worse...), but it could also be an additive factor in raising ocean temperatures. With catastrophic effects on the weather. As we know from history or projections for our future under Global Warming.

Yet again, knowing more about this phenomenon could allow us to control it in the future. If we survive long enough as a viable specie.

The technology for long-term survival is the subject of this book and also my forthcoming book *The Conquest of Ignorance and Human Survival.*

Manmade Disasters

General

Other than a nuclear holocaust or biological warfare the main one on offer at the moment is the effects of Global Warming.

Those who were in the forefront of thought on this subject have known about this for at least a decade. But as is common with people who have new and correct knowledge about events that will happen in the future, they were ignored.

Much as Homer Lea's prediction of the attack on Pearl Harbor and Hitler's rise to power.[8] Predicted up to two decades before the actual events. With extensive detail, including invasion maps. In his *The Valor of Ignorance* and *The Day of the Saxon* respectively.

Global warming and its effects

I just finished glancing through two tomes on Global Warming written by our (if you are an American) protectors. By American bureaucrats. One is "Our Changing Planet," by the Climate Change Science Program (CCSP) and the Subcommittee on Global Change Research (SGCR). The report has what might be termed a "seal of approval" as the report is a "Supplement to the President's Budget for Fiscal year 2006". Thus vouching for the efficacy, authenticity and other such things.

The other is the U. S. Environmental Protection Agency (EPA) Global Warming Report. Last modified October 2nd,

8 Homer Lea (1876-1912). An American hunchback,, he became a general in the army of Sun Yat-sen. And wrote *The Valor of Ignorance* and *The Day of the Saxon.*

2002.

These reports are 224 and 11,788 pages in length respectively. Beautifully done, attesting to the tens of millions of dollars probably required to create the data and subsequent reports.

Unfortunately though, I see no guarantee as provided with most private sector products. No real mention of efficacy and fitness for purpose, etc. Good move on their part, actually.

If they were not on a serious subject that means life or death for us and our specie they would be laughable. They have a three-pronged "solution". I am not making this up, this is serious. It is what they say. They propose:

1) Raise the beaches.
2) Erect dikes and levees.
3) Build pumping stations.

I can only assume that they know of some way to "jack up the beaches"? And to protect these with dikes and levees similar to what "protected" New Orleans? And the only place I think to pump the water is back into the ocean that is rising? You don't want to pump salt water onto land that you intend to use for anything other than "salt water fish farming" or mining of salt.

But you should be cheered to know that two experienced stalwarts will be handling the operation. FEMA and the "corpse of the engineer".

You don't need to take my word for this. You can buy the report from Amazon.com for $19.95. The title is "Planetary Emergency: Global Warming and the Future of Planet Earth, Climate Change Research Data, Greenhouse Gas Emissions, Carbon Dioxide Pollution." The ISBN is: 1-4220-0624-7. Published by the US Government. It comes on a CD-ROM.

Result of Warming

General

There will actually be many, many effects. The ones I will

list are probably only a small percentage of what will happen. Although I may be able to think of most of the major categories of things. And "one thing will lead to another". In some cases many other things.

A temperature change does not only mean that. The need to wear more or less clothes. It may dry out the land and dry up rivers. Animals will move as their habitat changes. And the hunters and those hunted may also move. If they do not, then it spells trouble for the specie. Bacteria and parasites will move as the host moves. But then to be replaced by others, which may represent a new threat to the original habitat.

Disease will move with the "herd" of bacteria and animals, birds, reptiles, insects, etc. Places that had never seen malaria will be plagued by it. Those that were plagued by it will be happy it is now gone. But perhaps not happy with its replacement, about which they know little.

Everything from small plants and grasses to giant trees will be at risk. And while one type of plant might be replaced by another, giant trees do not grow so quickly. And may not grow at all. And of course the plants are another factor in the movement of everything from insects to large animals and birds. And man. Many of the plants are eaten by human beings. So we get at least a "double hit" on this.

The rising temperature will also effect the sea. Which effects everything from the marine life to storms. Besides the current depletion of some fish stocks because of over-fishing, we may find that some simply cannot survive and will be gone. Again, they may take many other species with them. You must see that this is a cumulative, cascading, progressive thing?

The Tragedy

The tragedy is that we did not think in the long term. We did not use an appropriate epistemological tool to measure the rightness of things we did. And still do not. Maybe our short term thinking is best exemplified by the financial world. Rather than think of investment results in terms of at least decades we went to yearly, Then quarterly and monthly. Now sometimes it seems almost hourly. Buying or selling to make a fraction of cent in a fraction of an hour.

Meanwhile, we ignore the fact that pension, healthcare, education and other major societal systems do not work. We make war on whims, killing hundreds of thousands of people in small wars. Millions in larger ones.

We have embraced the concept of the "rule of law" to the point that now it is beginning to seem that unless something is specifically permitted it is not allowed. I was astounded to read that in England a woman may not "donate" her eggs to herself for possible future use. In some so-called "developing countries" people do not do certain things in business because there has been no legal framework created to cover it.

I think they refer to this as "enabling" legislation? A bit like a wooden puppet that needs to be enabled by the puppet master. Until the strings are attached—the control strings—the puppet can do nothing. Until the "control strings"—the political laws are created—the individual can do nothing. They are helpless without their political master.

Need I say, this makes a mockery of the concept of a thinking being?

The tragedy is that without individual freedom a thinking being does not think. A bit like driving a car. A person should really look at least hundreds of meters in front of the car to see what is in store for you. And use our natural instincts. Most animals will avoid something that does not smell good. Or perhaps feel good, look good. Most people recoil from the black smoke of a Diesel engine. Or try not to be near a belching smoke stake. The fact that this does not smell good should have told us something. At the very least provoked "wonder". To wonder if this might not be good for us in the long term. There it is again. That concept: the long term.

One time I had a very good boss. He was boss, friend, teacher. One of the first things he told me when I started working for him was "… if I walk by your office and happen to see you with your feet up on the desk and eyes closed, I will believe that you are taking time out to think. To think if what is happening is right. Or if there might be a better way to do something".

We know there has been climate change in the past. And sometimes very rapidly and very disastrously. Or at least it would have been for us if we had been there. So we are not alone in being the cause of climate change. But if we had thought about what was happening, we might actually have been able to stop any natural trend and to preserve the planet as it is now. Yes, this would require learning things we do not know now, but this is the story of our technological progress.

And we may still be able to do this. But time is running out in which to do it. And we must change our societal system before we can hope to solve the problems related to Global Warming. Change from a societal system based on coercion to one of total individual freedom. If we do not, we will perish regardless of Global Warming. With Global Warming sooner.

Maybe this almost sounds as if it should go to the end of the book? I think what I am doing is trying to give you something to think about as you read about our ascent as a specie and our descent. Why our present trend is toward the dustbin of history and what we need to do to prevent that.

Chapter 3

Ascent of Man

Synopsis

And such an ascent it is! In 3.65 billion years, going from a single cell to a thinking being with tens of billions of cells. A journey with dead-end roads and at least once coming perilously close to extinction. As we are now.

We can learn a lesson from this ascent. It is the competent individuals that survived. And created an increasingly competent specie. Something we have consciously tried to reverse by a misunderstanding of equality under the laws of nature.

Which says that each example of Homo sapiens is unique and different from all others. That it is this uniqueness, natural freedom and associated responsibility for action that ensures survival.

Background

With credit to my great friend and teacher, Jacob Bronowski (1908-1974) for the title of this chapter. I do not intend to imply that we knew each other personally as we did not. I say that in the sense that he was my teacher because I learned many new, correct and interesting things from him. And anyone who is your teacher must also be your friend, yes? The fact that he is dead is very sad, but does not prevent him from being my friend and teacher.

I watched his series—*The Ascent of Man*—on BBC TV when I was living and working in Holland. And I was there and watching the series when I head the news of his death. At first I could not believe it. I thought they must be speaking of someone else. Maybe I did not understand the Dutch language correctly.

For someone from whom I had received such value in life I found it hard to accept. I think to this day—about thirty years later—I can still hear his voice. And now, reminded again of him, I want to get both his book and DVD. I had the book, but it must now be in storage somewhere in my "techno-Gypsy" world.

I wish Bronowski could come back now and see the new information that has been uncovered and the new knowledge created in the past thirty years. I think at that time the earliest Homo found was about four million years? Now we are pushing seven. But the worst is that he—like I—would be appalled at our descent.

I am using "man" in the same sense Bronowski did, in the sense of the genus Homo and specie Homo sapiens. Which includes the female of the specie.

I am not against a language evolving, but the fact that avoidance of the use of "man" and "mankind" is for the wrong reasons. Thus I do not accept it. Any more than I accept "chairwoman" or, worse, "chairperson." Why not change "woman" to "woperson"? And "women" to "wopeople"? As I will point out in a later chapter there are much, much more serious problems around than that as exemplified by this nonsense.

This dispute is of about as much value to society as that related to "master" and "slave". Often this designation is used in technology. For example, as in "master disk" and "slave disk". Yet in the American political entity of California it recently became a "political" issue. And I believe such designation is now banned.

Although I do agree that the implementation of the concept in society should not occur. And politicians—the master—should not be able to make me their slave. Yet that is the way it is in politics, yes? The "acid test" of the existence of slavery is whether as an individual I can do anything I wish that is not immoral. If not, then someone is my master and I am their slave. Please notice I say "moral", and not "legal". As the latter concept has no place in a society of thinking beings.

It is the height of ignorance and impertinence for a politician to say they are "granting" people freedom. That exists naturally, They can only take it away, not grant it.

The natural condition of thinking begins—both male and female—is to be free. To have individual freedom. It is only through our adoption and maintenance of a societal system based on coercion—the system of politics—that this freedom is taken away. In a world of individual freedom there would be only one class of individual: Homo sapiens. And whether of one religious, cultural or ethnic group would be irrelevant.

There would only be two classes of people: those who act morally and those who do not. And it would be proper to "discriminate" against those who act immorally and to be prejudiced against them. Yet not to the extent of taking any action against them other than in their access to services or products provided by an individual. This will be discussed more fully later in the book.

Also, for Homo sapiens there is only one border on this planet. And that is the one at about 5,500 meters or about 18,000 feet. Beyond which we can not survive without artificial support. And even at that level we should not stay for long. Political borders are irrational and immoral.

What an ascent we have had. It is quite possible the intelligent species in other star systems did not go through the same long process. Interrupted by mass extinctions and climate changing threshold events. Although some may have had an even longer and more arduous journey. But 3.65 billion years is long enough, yes? And perhaps 7 million years for our direct, hominid-type ancestors?

While the 7 million years seems long compared to our present, ridiculously short life spans, in geological terms it is an eye-blink. And the essentially one hundred years we have been using fossil fuels seems very insignificant.

I would not dare attempt to tell you about the process of our ascent. I do not know it well enough and there are many good books and sources on the Internet for the details.

Rather, let me simply provide an overview with some important steps or points along the way. Of our path to become Homo sapiens sapiens.[9] Although our chosen societal system insists we must dispense with the "sapiens" part.

Let me make some generally accepted, representative assumptions:

- Assume that *life* began 3.65 billion years ago.
- That the first *hominids* appeared about 7 million years ago.[10]
- *Australopithecus afarensis* was here from about 4 to 2.7 million years ago. A small-brained ape that in 1.3 million years did not discover tools.[11]
- An early *genus Homo,* in the form of *Homo habilis* appeared 2.2 million years ago.
- *Homo erectus* appeared 2 million years ago.
- *Archaic Homo sapiens* 750,000 years ago.
- *Homo sapiens* about 400,000 years ago.
- *Homo sapiens sapiens* [us] 130,000 years ago.[12]

What *really* changed during the first four million years? In a sense, nothing. And even until about 1700 not much had changed. Imagine that!? Essentially it took us at least four million years (and you could say seven, even eight million years) to acquire even the beginnings of technology. Meaning a level at which new concepts in technology would appear within a few years rather than thousands of years.

---

9 There is a very interesting web site at: <http://www.humboldt.edu/~mrc1/main.shtml>

10 I had originally written 4.4 million years. Then I changed it to 6 million. But I had already read 7 million years. Now I see eight?! But it now seems moot. As if the present trend continues we will be extinct before the end of this century…

11 Every time I read and edit this I have problems. New information, pushing Hominids farther back in time. Interesting new information appears at: <www.fieldmuseum.org/evolvingplanet/hominids_1.asp>

12 There is a very interesting and succinct document illustrating much of the preceding : <www.wsu.edu/gened/learn-modules/top_longfor/lfopen-index.html>

Some have said that it took seven years to build the World Trade Towers and only minutes to destroy them. This is, like most thoughts and statements of a societal nature, only a partial truth.

It really took us up to 7 million years to build them. All but the last ten or so learning *how* to build them. And, while the act of bringing them down may have taken only minutes, it also took 7 million years to acquire the technology to do that.

The following few things, some of which occurred thousands of years apart, were of crucial importance in our development.

- Some animals were domesticated 15,000 years ago.
- Certain metals were created 12,000 years ago.
- Man learned to grow crops 11,000 years ago.
- Fired-clay pottery was created 9,000 years ago.
- Egyptians used crop irrigation 7,000 years ago.
- Cattle are domesticated and the wheel is invented 5-6,000 years ago.[13]
- Cuneiform writing originates in Sumer 5,000 years ago.
- With Aristarchus of Samos, after whom our planet is named, man begins to learn something about the shape and size of his planet 2,300 years ago.[14]
- Gutenberg invents moveable-type printing 550 years ago.

The last event is crucial. As the creation of new and useful knowledge was accelerating. And there needed to be some way to both preserve and to disseminate this. Single individuals writing it on paper-like substances was not good enough

---

[13] Probably invented many times in many places over a long period of time. Long by our life span. I wish I could somehow communicate with one of those individuals. To understand the moment of inspiration.

[14] It is commonly known as "Earth," although the proper name is *Aristarchus*. Just as the proper name for our own local moon is *Oberth*, not "Moon".

to handle the burgeoning "information explosion".

Even after Newton's publication of *Principia mathematica* in 1787 it took nearly a hundred years for this to permeate society and start the *Industrial Revolution.* And it was the knowledge contained in *Principia* that was the proximate cause of this.

Of course part of the problem was because it was originally published in Latin. Many of the Seekers and Creators of new and useful knowledge and products were not learning Latin.[15]

And it was nearly seventy-five years before the great Émilie du Châtelet (1706-1749) would translate *Principia* into the vernacular, in this case French. Making the knowledge available to a much wider audience.

If women are looking for one of their kind who was important in history they need look no further. She is one of the most important people who ever lived. Vastly more important than all of the politicians who ever lived. Whether of the civil, military or religious branch of politics.

The fact she was also a bit of a courtesan and the mistress of Voltaire only adds "color". She was also one of the most brilliant mathematicians of her age. Do a search on the Internet. She was a fascinating, wonderful person.

Dawn of Civilization

They say that "History Begins at Sumer", with credit to the author and book of the same name.[16] This may be true in the sense of being able to record what happened before. And in the sense of writing, making new thinking available to create a better future for Man.

Yet the approximate 3,000 year Egyptian civilization really left us with nothing of long-term value to the survival of the specie. Except for magnificent stone monuments to

---

15 With credit to Daniel J. Boorstin for the concept of *the Seekers* and *the Creators*, from his wonderful books of the same name.

16 Samuel Noah Kramer, *History Beings at Sumer*, (Doubleday, Garden City, 1959).

employ the archeologists of the future.

It was left to the brief time of the Greeks to begin to create new and useful knowledge.

The knowledge with which we could begin to create technology probably started with the Greeks. When we began to learn more about our planet and its geography. Its probable size, shape and place in the universe. And about mathematics and medicine. About what we might call "general science". The foundation upon which technology could be created.

There were the great thinkers such as Archimedes, Aristarchus of Samos, Hero of Alexandria, Hippocrates and Ptolemy. Even his "defective" planetary system was a necessary first start to knowledge in that area. A "first approximation". And many others who laid the foundations upon which Brahe, Kepler, Galileo and Newton would build.

At the same time we had the rise of the "great" serial killers, such as Alexander of Macedon. The most "successful" politicians of their age. Successful at what politicians do best: controlling and destroying property and people.

Well, isn't that their legacy? Name one thing of long term benefit to Homo sapiens created by a politician. Now as I edit this months later I still have not thought of anything. If you think of something, let me know using the form at: <universal-salvage.org.uk/mail.php>.

This knowledge creation process of the Greeks was interrupted by the Romans. Who concentrated on conquering people and building aqueducts, bridges, and magnificent buildings.

There were some history books. But they had nothing to write about except wars and conquering peoples. They did create some magnificent aqueducts and good irrigation systems. But as far as I know they elaborated no laws of nature. Produced no great and basic principles of science.

Yes, the arch is useful. But it would have been vastly more useful if they had understood that this was based on thought. And the freer the individual is the higher the quality and quantity of thought.

There were Hero of Alexandria, Ptolemy and Galen. But they were not Romans.

The Romans are generally credited with creating the body of unnatural law—political laws—to administer the activities of the ever-increasing number of people. They did this of course. But what people today do not seem to realize is that this institutionalizes the concept of societies based on coercion and fraud. And rather than solve problems it creates them. So they also left us nothing of long term value except archeological things. But more on that later.

Please do not stop reading because you do not like what I say. Instead read and think. You can learn something of value. Failure to continue will result first in the inability to solve the problems of our societal structure. And then those of Global Warming. With the result we will perish as a specie.

If you want to take a negative approach read with the idea that you might discover the flaw in what I say, that is okay. And try to expose me as a fraud. Not possible. But please continue anyway…?

The Roman obsession with dictators and totalitarianism seemed to put a hold on thought and the creation of new and useful knowledge. Their emphasis on a consumer-oriented society rather than a producer-oriented society. A situation that under the current American ruler Bush is returning with vigor. With its need for entertaining the masses to keep them happy. Something the Romans learned—almost invented—but it did not help them any more than it will help this era.

Probably the greatest man of the Roman Era—Publius Cornelius Scipio—has been almost totally ignored or misunderstood.[17] Although part of his achievement was political that part did at least involve self-defense. With the result that Rome became the "master of the Mediterranean" and the

---

17 Publius Cornelius Scipio (236-184 bce). Effectively the savior of Rome and then creator of the Roman empire. After Scipio it no longer needed creating, but could be destroyed.

Roman Ruler the ruler of "the known world".

Which resulted in the fact that Europeans and thus North Americans have an Occidental and Christian heritage. Rather than Oriental and Muslim. Which would almost certainly have been the case if Carthage had won the Second or Third Punic War. Think of that: if Hannibal had won, we would most likely be Oriental and Muslim.

Yet Scipio's greatest achievement seems totally unknown to anyone except B. H. Liddell Hart, Galambos and those who listened to and understood what Galambos said.[18] The achievement was Scipio's statement that "No man should rule other men. And certainly not for a lifetime". It was Galambos that brought this to my attention.

This was his statement upon being offered the rulership of the Roman empire after his victory over "the greatest general of all time", Hannibal. Who in that light becomes less great, yes?

Scipio was then only thirty-five years old. He was the most popular man in the Roman Empire. As well he should have been. After the death of his father, with Hannibal on the way to the very gates of Rome, no one wanted the job of taking on Hannibal. Even with the possibility and necessity to save Rome. The Senate only reluctantly accepted the militarily unknown and totally untried young (24) Scipio to command the armies of Rome against the most respected and feared general in the world.

At thirty-five Scipio understood what no one then or since has understood. Except for Galambos. That societies based on coercion can not and do not survive. Verified about six hundred years later when Rome collapsed as an Empire.

The true cause being lack of individual freedom. A statement I know you will disagree with. But again you must please *read* and *think* to understand why this is true.

The collapse of Rome and rise of Christianity—which oc-

---

18 B. H. Liddell Hart, *Scipio Africanus: A Greater Than Napoleon*, (Da Capo Press, New York, 1994).

curred at almost the same time historically speaking—ushered in the Dark Ages in Europe.

Please understand that when I am making the statement about the "known world" above that I do recognize the enormous achievement of individuals in China and elsewhere. Such as Persia.

But the fact remains that the ancient Chinese and Persian civilizations had—sadly—almost no effect on our modern technological world.

The Dark Ages

And they were truly dark. At least according to traditional history. And in the debilitating effects on mankind for more than a thousand years.

The only bright spot—ignored by "traditional" history and historians—is the magnificent Saracen or Arab civilization. That lasted from about 800 to 1500.

You can read about this in Henry Grady Weaver's *The Mainspring of Human Progress.* In Rose Wilder Lane's *The Discovery of Freedom.* And in my own books *Cowardly New World, Cracking the Bush Code: The Paradox of the Gospel of Bush.* And *The Conquest of Ignorance and Human Survival.* These latter two books should appear in 2007.

There is more information on all of these in the bibliography. And at <universal-salvage.org.uk/book_list.php>

The problem began at the Council of Nicaea in 325 ce, when the civil and religious branches of politics joined forces for their mutual benefit. And to the detriment of mankind. It was also here that by majority vote of mortal human beings the concept of the "trinity" was voted into being. The concept of the "father, son, and the holy ghost".

The loser was Arius, who believed that Jesus was of a *similar* substance as God (homoiousios) but not of *one* substance (homoousios). Only an "i" separated them. In a truly democratic fashion, by majority vote of mortal men, Arius lost. His losing group then became known as the *Arian Heresy.* And Christianity—at least the Catholic branch—

effectively had three gods. Or a single god with multiple personalities?

Arius was later "resurrected" or un-excommunicated. But sadly, on the way to the ceremony he stopped to do some shopping. And in the heat of Constantinople and excitement he suffered a fatal heart arrack.

In politics, and especially the religious branch, losers are often branded as heretics. In the civil branch of politics it may be as rebels, hooligans and other such things.

In the military branch of politics it may be as terrorists, insurgents or fighters. Although the civil branch of politics may also employ these designations.

All these pseudo-definitions and failed attempts at semantic precision result in a very confusing situation. Especially for a simple-minded and unimaginative person such as myself. Definitions should reflect reality and be unchanging. Not depending on friend or foe, ruler or slave.

The losers in political contests become the opposition, "loyal" or otherwise, with their actions sometimes being "illegal". Until they have replaced the current regime, in which case the tables are now turned. The actions of the former opposition are now legal, while those of the new opposition are illegal since they—having been supplanted by the previous opposition—now determine what is legal. And the cycle repeats itself.

I get the impression politics is a bit like a flightless bird? With left wings and right wings. Then "left of center" and "right of center" factions. It is a very unstable bird. Almost as it if is intoxicated. Sometimes going down "the middle of the road", other times not. If you are in the middle of the road you risk being run over by those constantly going from right to left and left to right. No wonder it has never been successful. Unlike that other un-flight-worthy bird the Dodo bird, politics has not had the decency to become extinct.

It was a stroke of genius and superior act of fraud when rulers allied themselves with God. When they assumed a title such as "His Christian Majesty". And let it be known that they had been appointed by God himself as his local representative. An aspiring ruler and associated accomplices might think twice about assassinating the representative of God…?

This set the stage for individuals to stop thinking. And to follow the dictates of Church and State.

This was the time of the First Inquisition. Of endless wars between petty princes. Of the first great Jihad. That of the Christian Crusades against an Arab and Muslim world of tolerance and relatively great modernity. Now George W. Bush, *Dour Leader*, has instituted the Second Inquisition in what was supposed to be the "Citadel of Freedom". Where individuals are told to extol his virtues. Not much different than *Dear Leader*, yes? And those who make *Dour Leader* "look bad" can be placed under hour arrest.[19]

I still believe that *Dour Leade*r should meet with *Dear Leader*. Even at the price of renouncing his weapons of mass destruction. This may seem a heavy price, but there is the possibility of a priceless treasure The political technology to become an *Eternal President*. As is the father of *Dear Leader*, Kim Il Sung.

And there is clear evidence of this desire on the part of *Dour Leader*. His mention of the possible need to postpone elections in the event of acts of terrorism. Which has an increasingly broad definition. For a correct definition see the appendix.

Further, it is not unknown for American politicians to manufacture "trigger events". Such as was done to justify the start of the American War in Vietnam. And of course as Bush himself did to justify his invasion of Iraq. Manufacturing or adopting causes that did not exist. Because, as is well known, he had long-decided to go to war with Iraq. And only needed a reason he could sell to his acolytes.

He could suggest the legislative branch take an extended holiday. To go home and think about how to combat terrorism. And he could rule by decree, something he has been

19 <http://universal-salvage.org.uk/books/warming.htm#muzzle>

practicing for years. After all, what is it when he states his disagreement with laws and intention to interpret them as he wishes?

Back to history. Before about 1100 in Jerusalem and other great Arab cities, Muslims, Jews and all others lived together in peace. As they did in Spain. The Jews were only persecuted and forced out of Spain when the *Christians* took control of the southern half of Spain from Muslims that had allowed all peoples to live there in peace and freedom.

The darkness in Europe—as contrasted to the light in the Saracen (Arab) World—only began to change with and because of the translation of the Saracen texts from Arabic into Latin. And the movement of these translated texts into Europe via France, southern Italy and Venice.

A prime mover in this translation was *Alfonso el Sabio* [Alfonso the Wise] (1221-1284), King of Castile (1252-1284). By the late 12th century the archbishop of Paris had a library of *four books*. Rather large for the time in Europe, but not available to the public. The *public* library in Cordoba—an Arab and Muslim library—possessed *32,000 volumes*. Open to scholars of any nationality or faith.

It was a time of great Arab advances in medicine, surgery, mathematics, astronomy, astrology. Greek writers of works lost entirely to Western Europe had first been translated into Arabic. And for centuries the Arab scientists had added to them. Then in the 1200s these began to be translated into Latin. With Alfonso employing an "army" of translators. Himself doing some of the translating.

He was himself a scholar, writer and composer. He composed some beautiful music. Among which were a series of sacred songs in honor of the Virgin Mary (the *Cantigas de Santa Maria* and *Cantigas de Toledo*) as well as other works.

He translated Arabic texts into Latin, including an account of the life of Alexander of Macedon. And commissioned the translation of many others originally in Greek. He wrote extensively on astrological subjects, drawing up tables of houses. He was a patron of the arts and scholarship, especially esoteric scholarship. He patronized the works of Jews, Christians, and Moslems equally.

Thus it was that much Greek philosophy, science and mathematics—preserved in Arabic—was translated into Latin and found its way into Western Europe through France, Venice and Arabic Sicily to Italy. The Venetians were learning anatomy and practicing dissection at a time when this was prohibited in the rest of Europe, using the anatomical texts originally written by the Saracens.

Christians, Jews and Muslims managed to live together in Muslim Spain and flourished in the court of Alfonso the Wise. But His Holiness the Pope was not pleased. Nor were the Christian nobility of Castile. Such flagrant tolerance was not to be allowed. Attempts were made to excommunicate and depose Alfonso, but he survived. Imagine that? Seeking to punish someone for tolerance. But, as they say "God works in mysterious ways"...

I have a special feeling for Alfonso el Sabio. For the preceding and because of something he said:

"Had I been present at the creation I could have offered some useful hints for a better ordering of the universe".

A great insight into astronomical knowledge of the past and his time. Cheeky fellow, yes? Some might say blasphemous. But how true. And in current American vernacular (if I am not behind the times too much), "Go for it, Alfonso"!

The Jews—and the Arabs who had accepted them as equals—were expelled from Spain when Granada fell to the Christian forces in 1492. Along with all Muslims who refused to be baptized Christian. Tolerance was replaced by the Inquisition. And the skill of Jewish and Arab craftsmen by the plunder and exploitation of the Americas.

Remember, it is Spanish *Christians* who practiced genocide on a mass scale in the Americas. Creating what we would now call "crimes against humanity". Which they were. They were brutal, deceitful and barbaric invaders. And European *Christians* who did the same thing with the native peoples of the American colonies. And as I mention elsewhere here, it was also European Christians who invaded the peaceful Holy Lands with their "Crusades". Traveling 10,000 kms in four yeas to bring death and destruction to a peaceful and tolerant Muslim world.

Further, now that I think more about it, we should also remember that it was European Christians—mainly British and French—who colonized and constantly gave pieces of the Arab and Muslim Middle East to each other from the 1800s into the 1900s. Forgetting there were many different peoples there trying to protect their own lands. And we wonder why they are sometimes resentful of the "West"?

The universe is comprehensible

Westerners should realize and be thankful for the fact that it was the knowledge of the Arab and Muslim world that created the *Renaissance* in Europe. Including beginning thoughts on individual freedom. And eventually led to the *Industrial Revolution.* The people of Israel should give thanks for this to the Arab and Muslim world. Rather than keeping millions of Arabs in the world's largest concentration camp.

As I often say, little different than that run by the Nazis. Missing mostly the ovens. Odd for the victims of the Nazis treating others as the Nazis treated them. Or is it more than odd? Perhaps criminal?

The Israelis have substituted American 500 pound bombs and laser-guided missiles for the ovens. More efficient, actually. Much higher temperature than the ovens, plus they have the feature of the explosion factor. Even Bush's cluster bombs, to ensure greater "collateral damage". Often for years to come.

In Afghanistan, Iraq, Lebanon and Palestine you might ask a question of someone whose house and family was just destroyed by one of these American export products: was or was this not a weapon of mass destruction?

How massive, how much destruction needed? Kill more than two people? More than a hundred? One child or ten children? How many women to qualify for this brand rating?

Actually this would not seem to have anything to do with the Ascent of Man, yes? More to do with descent. But in the "stream of consciousness" it just seemed to fit for me. If not for you, please still think about it…?

From the end of the Greek civilization to the late 1600s there was the feeling the world was incomprehensible. That it was not possible to acquire new and useful knowledge. Only to recycle and repackage the old fallacies. And although this would change in technology, it would not and has not changed in that knowledge related to our social structures.

But there were some who did think. Who, as the great Hungarian physicist Albert Szent-Gyürgyi von Nagryapolt said, engaged in *discovery*:

"Discovery consists in seeing what everyone else has seen,
and thinking what no one else has thought."

Those such as Tycho Brahe (1546-1601), Giordano Bruno (1548-1600), Johannes Kepler (1571-1630), Galileo Galilei (1564-1642).

Some names may not be familiar to everyone. And most will say "Ah, but these were just scientists. And the knowledge they created benefited only other scientists". Yes, their knowledge did not directly benefit society in general. Not in the short term. Although very much so in the long term.

It is necessary to create "fundamental" knowledge or "basic science" before you have the understanding to create technology. Which does have the potential to benefit everyone. And in primitive societies based on politics and coercion (one and the same thing) it is only technology that brings benefits to society. Politics brings no benefits to anyone. It is a valueless, problem-creating mechanism.

The unfortunate Bruno, and to a lesser extent Galileo, are stark examples of the truth of this. And of the problem that plagued the *Age of the Inquisition.* Bruno had the imagination and the audacity to suggest that our Sun—properly named Copernicus—might be one of only countless stars. Of an essentially infinite universe. Without the Earth—and thus Pope and Church—at the center of all life.

Bruno said:

"Innumerable suns exist; innumerable earths revolve around these suns in a manner similar to the way the seven planets revolve around our sun. Living beings inhabit these worlds."

For this, on the beautifully named and tragically used Campo dei Fiori in Rome he was burned at the stake on February 17, 1600.[20] On the orders of a Christian.

But not before he said "I await your sentence with less fear than you pass it. The time will come when all will see as I see…"

Such blasphemy and heresy. Anyone who would say this might say that Homo sapiens—me for example—descended from apes. And that my grandmother 4,000 times removed was black and ate bugs. Yuk. All but impossible for a white man who eats McDonald's hamburgers to accept? No, actually.

I think it is absolutely wonderful that I could have had such an ancestor. Who was fit enough to survive in that world. And whose genes were so good that her descendents were able to withstand the rigors of the following 80,000 years. Including the eruption of the Supervolcano at Lake Toba only five thousand years later. The possible rigors of the ice age, and the journey up through North Africa, the Middle East to become part of the Roman Empire?

To those who want to say they are Hispanic Americans, German Americans, Black Americans, I would lie to say that I am *Homo sapiens sapiens*. And a "time traveler"?

Galileo was in a sense luckier than Bruno. A luck helped by his willingness to recant. And rather then being burned at the stake he had the agony of living under house arrest for the rest of his life. For the "crime" of creating new and useful knowledge that did not please the ruler.

Just as Bush placed the climatologist John Hansen under virtual house arrest for daring to create new and useful knowledge for Homo sapiens. Which did not agree with the

---

20 I think this might mean "Field of Flowers" in English?

"gospel according to Bush". And did not "make him look good".[21]

In the 1600s, to a large extent knowledge creation moved to England. Newton was born in the year Galileo died. Fittingly, although according to the Julian calendar and not our Clavian version, this prophet was also born on what is known in the Christian world as "Christmas day", December 25th. 1642.[22] Newton was a prophet also. For all those who seek new and useful knowledge.

Everyone knows that Newton created the laws of motion and gravity. And certain laws of optics. Invented a reflecting telescope.[23] And explained his laws of motion and gravity in *Principia mathematica.*

But these are not his most important accomplishment. Rather, it was his bold statement that "*the universe is comprehensible*". More than that, he also described the intellectual steps required to create new and useful knowledge. And of course to verify existing knowledge for correctness.

And when about seventy-five years had passed after the death of Newton and enough people understood what he had said, it was this that started the *Industrial Revolution.* And as Henry Grady Weaver (1889-1949) said in *The Mainspring of Human Progress*:

> *"Why did men walk and carry goods (and other men) on their straining backs for 6,000 years—then suddenly, on only a small*

---

21 <www.bbc.co.uk/sn/tvradio/programmes/horizon/dimming_trans.shtml>. Be sure you select & copy all of this URL.

22 This is normally and incorrectly referred to as the Gregorian Calendar. The only thing Pope Gregory XIII had to do with it was to impose it on others. It was actually created by the mathematician and astronomer Christopher Clavius (1538-1612)

23 Robert Hooke (1635-1703) also invented a reflecting telescope. Both are conceptually based on the same design as that of the Scottish astronomer James Gregory (1638-1675). Niccolo Zucchi (1586-1670) probably invented the first reflector telescope in 1616, but poor optics prevent it from being successful. I think it is accepted that both Newton and Hooke knew of Zucchi's work. And that their telescopes were improvements on Zucchi and not completely original.

part of the Earth's surface the forces of nature are being harnessed to do the bidding of the humblest citizen?"[24]

The answer is actually two things. One of course was *technology.* But the other was *individual freedom.* Both of which will be discussed in more detail later.

Technology

One useful definition of technology that occurs to me is "the application of basic knowledge to the solution of everyday problems".[25] The problem of not having light at night. The problem of not knowing how to fly. The problem of hauling heavy loads over long distances. The problem of having the plague or smallpox. The problem of not being able to afford a minstrel gallery in your house and musicians as full time household staff. So that you can have music when you wish.

Knowledge, like ourselves, is evolutionary. And accretionary. It is built like a brick house. Brick by brick. Bits of knowledge by more bits. But as Jules Henri Poincaré (1854-1912) said:

"Science is built up with facts, as a house is with stones.
But a collection of facts is no more a science
than a heap of stones is a house."[26]

In a sense that is what a Newton, Einstein, or Galambos does. They not only create new and useful knowledge but they *organize* knowledge. Through the creative process they organize it. They take "stones" created by a number of ideological antecedents. Think what it all means. Rearrange the stones, add a few new ones. Then create a new hypothesis and test it using the last step of the Scientific Method. When it passes it becomes a theory. Another stone or bit of knowl-

---

24 Henry Grady Weaver, *The Mainspring of Human Progress*, The Foundation for Economic Education (1997), p 4.

25 I am independent on that if not original. I have never seen it anywhere or heard it at any time. It occurred to me as I wondered what might be a suitable definition of *technology*.

26 Quoted from Bertrand Russell in preface to *Science and Method* {1913}

edge. Or something big enough and important enough to form the foundation for a new house.

Such as Newton's *Laws of Motion*. Einstein's *General Theory of Relativity*. Galambos' *Universal Laws of Volitional Behavior*.

As I edit this it occurs to me that perhaps here is part of the key to the failure of socialist school systems. The "free", public school systems of most of the world. Which require that students collect a huge number of stones. Such as the date of the Magna Carta. The date and location of Napoleon's battles. The value of Π. The formula for determining the area of a circle. For converting degrees Fahrenheit to Celsius. But not *how* to do these things. *How* the formula was derived.

Another definition for technology might be that it is "the application of the laws of nature to create a tangible product of value to individual thinking beings." With "value" meaning that it can add to the quality of life for one or more individuals without having a detrimental effect on the life of others.

Unlike politics, where the creation of each unnatural law favoring one or more individuals has a detrimental effect on the lives of many others. Which is the cause of chaos and disorder in societal life and a major reason we have not and can never solve any societal problems through politics.

A major problem with some of the technologies we have used is that they have had unforeseen, damaging side effects. We could not have foreseen the effects the burning of fossil fuels would have on our environment when we first started to use the associated technologies. The required knowledge of the laws of nature related to combustion, thermodynamics and climatology did not yet exist. Indeed, climatology itself may not yet have existed?

And when the knowledge was created it still took some time before we would understand how combustion effected our atmosphere.

Our nose might have told us something—although not directly—about the damage to our environment. Yet studies have shown that when animals—thinking or otherwise—

detect an unpleasant odor it often means that consuming the item causing it may be damaging to the animal. And that maybe we should find an alternative that did not involve the creation of pollution.

We should have thought about the effect of the black smoke coming from smokestacks. That the smoke was not a good thing. And some did. When I was a child in Los Angeles almost every home had an "incinerator" in the back yard. Which people used to burn some of their trash. And sometimes also used it for outdoor cooking. In about 1944-45 the city banned them. Part of their "regulatory" process.

There is however a very major problem with this coercive regulatory system. At least two problems. One is that people soon seek to do only "what the regulations say". And no more. The other is that it becomes a game. People do not like being told what to do so they contrive ways to "get around the regulations".

The Titanic is a good—or bad—example of the first problem. During construction a young engineer noted there were not enough life boats for the planned total of passengers and crew. And reported it to management.

He was told that the number "satisfied regulations". That did not satisfy him and he complained further. And was eventually told that his continued employment depended on dropping the subject. And many passengers and crew died because of state regulatory coercion.

Many people would say this was caused by "greed" to make a "profit". And is thus the problem of the "dirty businessman". In a political world "profit" has a bad image and "dirty" or "greedy" is almost a required adjective for "businessman".

As usual in things political this is a partial truth. But you must remember that it is *first* the regulation and control of individual life and *then* the actions to "get around this". To find a "loophole". And to resist the control.

I have thought of something that I think might qualify as a law of nature:

"There is a direct relationship between the attempt to eliminate uniqueness and the attempt to impose equality in volitional societies, and the level of frustration, violence and disorder that will ensue."

Without politics—a societal system based on coercion—war could not occur. If a rancher named Bush announced that he wanted to attack and destroy someone who had threatened his daddy and wanted hundreds of thousands of people to go to war against this person, most people would pay no attention. Who would give up their job for the chance to "kill or be killed" in pursuit of this objective?

In its basic form technology is merely knowledge. Knowledge that can be used to make life better for all thinking beings. With knowledge being amoral. It is neither moral nor immoral. It is only human action that can have a morality content. And of course a rationality component. But rationality is only for the account of the individual unless the action is also immoral.

Another very important aspect of technology is that it is offered in the "red market". The free and moral market. Meaning that those in the market have a choice. To buy or not to buy. To avail themselves of the product and its technology or not to do so.

The "products" of political states are offered in the "black market". The state-controlled market. The rules of which are that victims of the regime—commonly called *citizen*—must accept all offered products and at the price offered. No choice in product or price. Thus this is the true "black market". Black being the color of death and despair.[27]

What the smuggler is doing is moral but illegal. What the politician is doing is legal but immoral. As Galambos also said, "Politics is a system of false alternatives. Would you rather have a broken arm or a broken leg?" Would you rather pay your taxes or go to prison? The rational person wants

27 This thought and statement is with credit to Andrew J. Galambos. Who said "The black market is the moral market in an immoral society".

neither, but there are no other alternatives on offer.

As George Orwell said:

"In our time, political speech and writing are largely the defence of the indefensible."

How else would you refer to a "market" or system in which there has been war in every fourteen out of fifteen years for the past 3,000 years?[28] A market in which there has never been a solution to even one of our basic societal problems for 3,000 years. Well, has there been a solution to any problem?

In early 2006 I sent an Email to Naom Chomsky of MIT. Sort of a gadfly in technology and sociology. Asking to have a dialogue related to the fact that we had never solved even one of our societal problems. Telling him I knew why this was true and the solution.

He replied rather testily that we had in fact solved many of our societal problems. And were continuing to do so every day. I sent a reply expressing surprise and great interest in knowing about some of the problems solved. There was no answer. As of course no problems have been solved.

But hope springs eternal. Even if life does not. Surely now that the *ancienne régime* is out and the *nouveau régime* is in, problems will be solved, yes? And as they sing at a convention of the social democratic wing of the American Socialist Party, "Happy Days Are Hear Again".[29]

But as I say in other books, within hours of a decision there can be the feeling that a horrible mistake has been made. And "the candidate"—their selected political agent—will not solve the problems of each individual.

When you think about it—and you must think about it—it is not possible for one person to direct the lives of even two people and allow them to fulfill their individual hopes

---

28 Henry Grady Weaver, *The Mainspring of Human Progress* (The Foundation for Economic Education, 1999), 263

29 Americans like to believe they have a two-party system. And while organizationally they appear to be different, in reality they are the same. With the *Republican Socialist* and *Democratic Socialist* wings.

and dreams. Let alone doing this for millions or hundreds of millions of people.

And that in fact the concept of politics *must* create a system of *Government Without End.*[30] In which eventually everything will be either required or prohibited. And all individual thinking stops.

We have ascended physiologically. We have much larger brains than we did 7 million years ago. But have we ascended in a societal sense? Our tribes have become much larger. Such as the Great North American Tribal Confederacy (the USA). With at least tens of thousands of small tribal groups that do not really get along with each other.

They do not have real wars and kill each other. But with their tens of thousands of Tribal Chiefs and associated Witch Doctors (now called "spin doctors"), they still can not solve any of their mutual problems.

This could lead a person to believe there might be a problem in solving the societal problems that will arise from Global Warming and rising sea levels? And of understanding the cause of Global Warming and slowly arresting the process? Perhaps even reversing it?

Read on…

---

30 A situation I suddenly realized in the summer of 1956. In a time of relatively much greater individual freedom that—by the very nature of politics—this must change. With the end result being everything being either forbidden or required. Individual thought and action no longer needed or allowed.

Chapter 4

Descent of Man

Synopsis

Our descent probably began almost simultaneously with our ascent. We started as an intuitive animal responding to our environment. About which we knew little. And it is very difficult to make the transition to a thinking animal, increasingly able to control its environment.

We have not sat down and thought "we are a thinking animal. Each example of the genus and specie unique and different". And all that implies. That for the betterment of the specie, each individual must become increasingly competent and fit. To achieve "fitness" and survival for the specie. That the only way this can be done is by each individual assuming the responsibility for discharging the duties and obligations of individual thinking life.

We have thus accumulated an increasing number of fallacies related to societal life. Many with a high mystic content. And have not been able to get rid of the Tribal Chief and Witch Doctor syndromes.

This is what is propelling us in our descent…

Background

We have taken a step backwards from that of a thinking specie. Homo sapiens sapiens. We started to become thinking beings. Each example of Homo sapiens being unique and different from all others. With differing hopes and dreams for our life. And the associated abilities to implement our hopes and dreams. With this being a dynamic process, constantly changing throughout life.

A situation in which each individual took action based on

their thoughts. And was then solely responsible for the results. A situation in which individuals assumed responsibility for rationally and morally discharging the obligations and responsibilities of individual thinking life.

For a thinking specie this freedom of action and responsibility for action increases competence. Which can also be interpreted as "fitness" for the individual. Collectively, it is fitness for the specie. Essential for an increasing individual quality of life. And for the survival of the specie.

Yet as I say this, it is clear to me that we only rarely reached a modicum of individual freedom. So it never had a chance to become "institutionalized".

I think the "descent of man" started very soon after the ascent started. In an early, primitive society of thinking beings, brawn and physical abilities would naturally overshadow rational, long-term thinking abilities and thought. Although the witch doctor—the thinking branch of the specie—could be very powerful. But in a sense acting more in response to the actions of the tribal chief. The political ruler. And not exercising a proactive role.

Rather than opposing the tribal chief, the witch doctor allied himself with the tribal chief. Only challenging the tribal chief enough to create an identity for himself within a secure position.

Yet both positions had a major problem. The same as with "modern" political societies. They sought to stamp out individual life. To limit, even eliminate the ability of their victims to be unique and different. To implement their unique hopes and dreams for their life.

Which, as we should know, generates depression, irritation and anger. And the search for "loopholes" that is a major cause of the disorder and chaos in society.

I warn you: this is for thinking people. Who wish to think of true results and not of fallacies and fables. Not for conservatives, conserving the fallacies of the past. If it contravenes cherished fallacies and belief in mysticism then you must think about it using true premises and valid and logical thought processes to create a new hypothesis. That you can then test for reality with what *really* happened in history.

Cause of the Descent

General

In the beginning simply the lack of knowledge. As it began to accumulate it was the failure to understand it or the inability and unwillingness to apply it. We missed crucial inputs all along the line. We started with the concept of the Tribal Chief and Witch Doctor, and simply refuse to get rid of the concept.

Following are two thoughts: one is that individuals represent the creation of new and useful knowledge. That could have improved the lives of individual thinking beings. Propelled us forward in our ascent. But when we did not understand the knowledge and use it properly it was part of our descent.

Yet we must understand it soon. And implement it. Else our decent will have its inevitable end in extinction. Clear and rational reasoning, with a test for morality on intended action forms the basis for our survival as a specie. Or indicates the reason this did not occur.

Civilizations & Epochs

General

For the descent we must think of *results*. Those things that did not improve the quality of individual thinking life. As it is Homo sapiens and their civilization that we are discussing. And those few things that did add to the quality of life. Add to our quest for knowledge.

I will assume that we started our descent in earnest with the beginnings of the great city-states.

Pre-Egyptian World

This covers the period from the first genus Homo to the start of the First Dynasty of the Egyptian empire around

3,000 bce, or about 5,000 years ago. By this time some animals had been domesticated, certain metals were known and man had started to grow crops. By then Egyptians had been using crop irrigation and writing had been invented at Sumer. The wheel had been invented about 1,000 years before.

The Sumerians did bequeath to us the concept of writing we now use. And as I said, there was the domestication of animals and some agricultural and metal knowledge. But without individual freedom and responsibility for action and a non-dependence on gods, there was nothing of great value. Other than the usual things to provide employment for future archeologists.

I do not want it to seem that I am dismissing 7 million years of human evolution in just a few words. The "mere" survival of the specie was an enormous achievement. Against the greatest of difficulties. I guess in a sense the question would seem to be why hundreds of thousands of years would pass—even millions—without the development of a tool. In a very real sense there was no change in life for 7 million years. Minus the five thousand years from this period to now?

And then, in the last one hundred years of the 5,000 years—and also of the 7 million years—we throw it all away.

Egyptian World

The Egyptian Empire and world started about 3,000 bce, or 5,000 years ago with the First Dynasty. And in a sense lasted even through Cleopatra as an empire, although somewhat diminished.

But in a societal sense nothing really had changed. The Tribal Chief was now called Pharaoh, with grander ideas about conquering and controlling other tribes and their people. And with the exception of the time of the Pharaoh Akhenaten, there was still a system of a multiplicity of gods. One for almost every purpose. The Witch Doctor had become the High Priest.

The Greeks

Some people say that Thales of Miletus (ca 635-543 bce) was the first great Greek scientists. Se we might say that the beginnings of scientific knowledge began about this time. The cause of this was the novel idea of individual freedom. Although Greece had a rigid class structure and people were generally only free within their class. And perhaps not to migrate upwards. Yet even this modicum of individual freedom created the environment in which people started to think seriously and in an organized fashion about their world. With the idea of learning more about it.

The Greeks created the beginnings—the foundation—for knowledge about out world. The foundation for technology that would develop in the future. They did this in geometry, mathematics, logic, medicine, astronomy. Even art and sculpture. This would overlap the ascendancy of the Roman Empire founded by Scipio in 202 bce, with his victory over Hannibal. After Scipio the empire did not need founding again. It only remained for it to be destroyed.

The great astronomer Ptolemy (85-165) lived three hundred years after Scipio. Well into the Roman era. We should not look down on "the Ptolemaic system". He did very well with the knowledge at hand, and his work was a necessary precursor for the astronomers of the Middle Ages. Who in the 1500s would use his work as the basis for their own great advances.

Although we must feel sad that the world forgot or lost the knowledge of Aristarchus of Samos. That the Earth-Moon system revolved around the Sun. And wonder how this knowledge was misplaced for thousands of years.

With their relative individual freedom the Greeks still had their gods. No great leap out of mysticism yet. Although Abraham had the concept of a single god 1,000-1,500 years before the Greek system. During the dying days of the Greeks and the start of the Roman civilization Christianity would appear. Also a single-god system.

The Romans

The ascendancy of the Romans meant the end of the individualism of the Greeks. The Romans had a burgeoning empire to administer and needed regimentation and obedience above all. The oath of allegiance to the emperor and then on down the line to the lowest soldier, bureaucrat and ordinary citizen. And of course the slaves, of which there were many. I think in this respect the Romans may have been worse than the Egyptians?

The Romans were low-level technocrats. They built roads, aqueducts, irrigations systems and some magnificent buildings. Often meant to impress the natives with their military and economic power. And as with American and other modern politicians, to reflect their own glory to the people. Although I think they phrase it another way.

The Romans also had a multiplicity of gods. And their lack of individual freedom did not provide time or incentive for individuals to wonder and speculate about their world in other than a political sense. I think the first great historians did appear here. Such as Plutarch. Although I also think they tended to write what the politicians wanted to hear.

Yet among these we seem to see some Greek names? Strabo, Diodorus Siculus. Mixed in with the Romans, Plutarch, Tacitus, Titus Livius, Dio Cassius and others.

The Romans are credited with providing us with a "system of laws". Which they certainly did. They institutionalized the concept of the ruler and ruling party faction, and those ruled. I would not call this a positive achievement. It put the nail in the coffin of individualism, at least in the "Western" world to come.

I believe it is accurate to say that the Roman empire died from a severe case of politics. Advanced socialism and too little individualism and rational thought about their world. There is an inverse relationship between the number and complexity of unnatural laws and the period of survival for a political entity. As the number and complexity of laws—the rigidity of the system—increases the civilization begins to die.

A societal "rigormortis"?

It took the Roman civilization nearly a thousand years to rise and fall. It took the Soviet Onion [sic] less than one hundred years to unpeel. It will probably be less for the European Onion [sic]. The key actually is keeping the people content. Making them love their servitude. The politicians in the USA excel in this. Which is why the USA outlasted the Soviet Onion. Although the USA is still only a little over 200 hundred years old, it is obvious it is in an advanced and terminal state of decay.

In 1946 Aldous Huxley said making slaves enjoy their servitude would be the great problem for world politicians.[31] While predicting they would succeed at creating a totalitarian world as long as they could keep their victims happy, as they have mostly done.

While the Roman civilization was expanding they could do this. As it expanded for hundreds of years. And when it died it left nothing but archeological ruins and the disastrous concept of unnatural laws.

The Saracens

I will not talk much of the great Saracen civilization in this chapter. I have said more about that in my forthcoming book *The Conquest of Ignorance and Human Survival.* This was a very bright light of individualism from about the end of the Roman empire until the late 1400s. It was essentially a single business empire within a large number of political entities.

There were great advances in the same areas as the Greeks had worked. As it was the Saracens (Arabs and Muslims) who copied many of the Greek texts into Arabic before they disappeared. And in the climate of individual freedom and religious tolerance that existed in the Saracen world the great Arab scientists added to the knowledge of the Greeks. Without the knowledge of the Arabs, Henry the Navigator

---

31 Aldous Huxley, foreword to *Brave New World* (Harper Collins Publishers, 1946), xiv-xv.

and Columbus would not have been able to make their voyages of discovery.

And in the 1200s to 1400s, before the Moors (Arabs in Spain) succumbed to the Christian invader, many of these were translated from Arabic into Latin. And it was the appearance of these texts in Europe that fueled the Renaissance. And then made possible the Industrial Revolution.

These texts first appeared in Arab Sicily and then in southern Italy and Venice. At a time when dissection was prohibited in Europe, the Venetians were conducting autopsies and greatly adding to anatomical knowledge.

I will discuss the Saracen civilization in the following chapter on the Natural Society.

The Dark Ages

And they were truly dark. I believe everyone who might read this book has images of the Inquisition in Europe. The burning of Bruno at the stake for saying that our planet Aristarchus was not the center of the universe. That Aristarchus revolved around the sun and there was an infinite world of stars and planets. And Galileo under house arrest, barely escaping death during the last part of his life.

The time from the end of the Roman empire until the 1500s in Europe was a time when people thought only of the past, and that new knowledge was not possible. The best was to re-interpret Aristotle to find his true view on things. People were actively discouraged from expressing new ideas and creating new knowledge.

And of course there was no thought of individual freedom. God and his representative—perhaps your local baron—controlled all aspects of life.

The Intellectual Revolution

Originally I had said "industrial" rather than "intellectual". But the latter is more true and basic. For it was the period when man began to think. There was enough individual freedom that there could be the beginnings of thought and wonder about our world. And except for isolated individuals, in Europe this occurred mostly in Britain. Not just England, but also in Ireland and Scotland. After Galileo died the focus shifted mostly to Britain. And while those who indulged in politics sometimes found themselves in prison or executed, if you weren't interested in what the régime was doing and not opposed to it, there was enough individual freedom so that there was a true "flowering" of thought. It was only in the late 1800s and early 1900s that this began to shift to the USA. I am tempted to use "America" rather than "USA", but those living in Central America and Latin America are Americans also.

Yet the individual freedom that had existed in the Saracen world was lost everywhere except in the American Colonies. And while for several hundred years—from about 1650-1850—it existed in the USA, it began to die there also. The vision that Thomas Paine and the other "founding fathers" had was not explicit, detailed and forceful enough to prevent the rise of the first democratic politicians. Then in the 20th century a very virulent socialism, bordering on communism. Very little that separates socialism from communism. Socialism is "merely" state-control of all aspects of life, and communism is the state actually professing to *own* everything. But of course political states can not own anything, as ownership is a moral concept. Whether the state "only" professes to control and not "own", there is no practical difference.

Beginning in the latter half of the 19th century in the USA, politics and politicians gained the upper hand in societal matters. The turning point was in Franklin Roosevelt's response to the "depression". His vast expansion of the bureaucracy, intruding into the lives of everyone. Which actually made things worse. It was only the Second World War that began to pull the USA out of the depression. Roosevelt's socialist actions only exacerbated the problems.

When the failed concept of democracy was converted to a religion in the USA in the later half of the 20th century there was a very rapid descent into non-viability for Homo sapiens.

It truly provided an "opiate for the masses". Leading to a failure of rational and original thought on societal matters that had not existed since the Inquisition. Missing only the "inquisition police", which *Dour Leader* (Bush) is now deploying during the resurgence of the Inquisition. Not seen since the ascendancy of the Nazis in Germany.

Summary

Essentially, in the 4,500 years from about 3,000 bce until 1500 ce—with a few short or isolated exceptions—no one even thought of individual freedom. Let alone implemented it. These exceptions were the brief and rather tentative effort of the Greeks, and the longer but isolated event of the Saracens.

Although we must remember that individual freedom is something you do not implement. It is the natural condition for a thinking specie. An individual politician can only take action to *restrict* individual freedom, not to create it.

The English kings reluctantly gave up a little bit of their power. But this was then assumed by Parliament and the prime minister. So as with the idea of kingship itself, there is only a change in name. *Kingship is Dead! Long Live Parliament!* With Parliament being more like a raucous committee advising a crownless monarch. The difference between a dictatorship and a monarchy and democracy is only cosmetic. It is still rule by coercion. And as I point out elsewhere here, in any democracy there must be and will be ever-*increasing* unnatural laws. And with that ever-*decreasing* individual freedom resulting in ever-increasing disorder and chaos.

I think I am correct in the following statement. And that it probably represents a law of nature.

> *There is an inverse relationship between the amount of individual freedom and the chaos and disorder that will result in a society.*

It is the *reduction in freedom that comes first* that causes the disorder. Then apparently requiring unnatural laws to try to stop the disorder. Which starts a situation that ends up with such disorder that the society self-destructs. The political alternative being a situation in which everyone just gives up thinking and acting on their thoughts and you have an *Orwellian society*. Such as in Orwell's *Nineteen Eight-Four*.

In the USA the "last bastion of freedom" has been converted to the "last bastion of democracy", and our fate is all but sealed.

The descent continues…

Religion

General

The civil branch of politics (the other two being the military and religious branches) produces an ever-decreasing quality of life. Offset only by increasing technology. Which is almost the same as the Roman recipe to pacify the masses: bread and circuses. In an advanced democracy it is welfare, entertainment and shopping. In the increasing climate of repression and depression—and with no individual control over life—the masses must have some outlet. Another world with its message of hope for a better future. And this of courses is religion. A *fabulous* world it is. With "fabulous" meaning "of or related to fables". A self-administered anesthesia?

Abraham (ca 1900 bce)

Probably many historians think of the pharaoh Akhenaten as the "first great monotheist". I know I did until just now (2006). At the end of my life rather than near the beginning. I am not a "good, practicing Christian". Which might be part of the reason.[32] Because I feel that perhaps many Christians do accept Abraham as being one of the major prophets? As Muslims consider Christ to be one of the early, great prophets.

---

32 Although born a Christian I am a Deist. As Thomas Paine said, "My mind is my church". And my "religion" is to try to use it properly. For rational and useful thought that has the potential to enrich my own life. And hopefully that of others.

I would criticize the peoples of the world for not beginning to accept the end of polytheism after Abraham. But perhaps that is not totally fair. People should think that if there needs to be a god that one is better than many. And that if a person is really God, they would not create sub-gods or rivals. Confusing and not economical. Yet there is the problem of communication. At the time of Abraham it could take decades, even centuries for new knowledge to circulate around the world.

And in fact the knowledge of Abraham and of Akhenaten never got to China and Asia until recent times? And history does not always record truly great inventions. With some of the same basic innovations being created by more than one person at different times in history?[33] And lastly, the conservatism of people in not being willing to accept new knowledge when it conflicts with what we believe to be true. The concept of "intellectual honesty".

Thus the world stumbled on its way for another five hundred years to Akhenaten. The person until now that I referred to as "the world's first great monotheist." A position I now acknowledge belongs to Abraham.

Akhenaten [Amenhotep IV] (ca 1393-1362 bce)

The concept of polytheism—a multiplicity of gods—was the cornerstone of early Egyptians and their empire. What I might call the first great "nation state". Although I think this is usually reserved for much later in history. The ancient Egyptians believed that it was their gods who favored them and allowed them to be successful in the conquering of other peoples. And the booty thus taken was usually delivered to the temples for the priests to guard for the gods. Giving them great financial power in addition to their spiritual and mystic power over their victims.

---

33 The concept of evolution was articulated completely independently in the 1800s by both Charles Darwin and Alfred Russell Wallace. With each acknowledging the other's independence.

The polytheists in ancient Egypt had a strong union. As if everyone had the label in their shirt. They were not about to let Akhenaten get away with the idea of one god. Cutting into their power and access to the treasury. And after less than twenty years of rule they murdered both he and his lovely wife Nefertiti. But left their step-son Tutankhamen to make the pilgrimage back to Thebes from Amarna with his young wife Ankhesenamun.[34] More of a flight actually, as would occur with Muhammad about 2,000 years later.

Monotheism provides for an individual God. And means that the mortal individual may communicate directly with God. Rather than dealing through an agent or broker who does the communicating with God. The agents being the priests. While the Jewish, Christian and Muslim faiths have their rabbi, priests, and mullahs, they are not quite the entrenched bureaucracy that exists with polytheism. With its many gods you need the priests to determine which god to communicate with and how. Which was sometimes by physically sacrificing your life to appease one of the gods.

Christ (ca 0)

Two thousand years ago there was no common requirement or practice to record births. And while this has become common in the USA and Europe, it is still not standard practice some places in the world. Because of this we are not completely sure of the birth date of the Prophet Jesus. Some believe it may have been several years before the time now used. As often happened in history, the date of a person's birth did not become important until they themselves had become famous. Or infamous. And by then it was often too late to now exactly when the birth did occur.

In my database of important people, which has no politi-

[34] Who was probably at least 2-3 years older than her husband. Sadly, their only two children were aborted or stillborn, leaving Tut with no heir when he died at about seventeen. Probably from the result of serious wounds in the chest area received in battle.

cians in it, I have often taken the liberty of assigning a birth date to individuals whose exact birth date is not known. I think this should be acceptable. As it is good to have a date to celebrate so there is no excuse not to do so. And while the date of death becomes known because they have become famous, this is not something to celebrate.

I find it difficult to accept the idea of any god who visits disasters upon those who worship the relevant god. Who continuously allows a never-ending stream of disasters for not only those who worship this god but for all others. As with our current world. It would appear that the Christian God is visiting disasters on the Muslim world and peoples. Although I am not sure that the reverse is true? To me it seems that Allah is less bent on revenge than Jesus and his god? If we are to believe the Bible. I am trying to prepare myself to step aside if I see the bolt of lightning coming in time…

As a born Christian (once only, and not again) I feel I can criticize my own religion. But am very reluctant to do so with another religion. Although the concept of religion in general, yes.

But it is not necessary or called for here. I have done that in the chapter on "The Origin and Purpose of Religion" in my book *Cowardly New World* and elsewhere.

The one thing I will say here is that I do not understand a god that will allow the continuous disorder, chaos and destruction that continues to exists in the world. And I do not accept that this is okay because God "operates in mysterious ways". And this is not something I need understand. This violates the basic concept of a thinking being. One who trends toward rational thought and takes actions accordingly.[35] And who is then individually and solely responsible

---

35 This is the proper definition of *intellectual*: "One who trends toward rational thought". This definition eliminates most people commonly thought to be intellectuals. And raises many not thought to be intellectuals to that position.

for the results. And if that action is immoral and is therefore a crime this is not acceptable behavior. It is not good for the individual in the long term. And the summation of individual behavior is our society, civilization and specie.

This event did not help mankind…?

Muhammad (570-632 ce)

The Prophet Muhammad is the great misunderstood individual of our time. Perhaps as much by his followers as with others. And as with the Christian faith there are internal problems. Christians have their Catholics, Protestants and more minor divisions. The Muslims have their Shiites, Sunnis and more minor divisions.

Oddly in a sense, both the Prophet Jesus and the Prophet Muhammad were individuals of peace. This is obvious in what I understand to have been the life of both. And as articulated in the Holy Qur'an. While the Christian bible—written hundreds of years after the death of Christ—tells an entirely different story.

The successors and followers of both prophets do not seem to have the quality of the prophets themselves. Muhammad did create the concept of the *Jihad.* But it is much misunderstood. As originally conceived by the Prophet Muhammad it was a matter internal to the individual. The struggle with the ego and the possibly morally impaired individual And not an attack on others.

And the Prophet Muhammad was specifically against the concept of revenge. When the invaders from Makkah were repelled in their invasion of Medina, Muhammad specifically told his followers there should be no pursuit and attack of the invaders and the town of Makkah. That it was a time for internal reflection, thankfulness and peace.

The Christians perverted the concept of *Jihad*—although perhaps unknowingly and unintentionally—with their Crusades. And created not by God, but by his sole authorized representative on the planet, Pope Urban II. An attack against those of a different religion which they did not understand.

And one under which those in the "Holy Land" had enjoyed centuries of religious freedom. Where Christian, Jews, Muslims and all others were worshiping freely and side-by-side. Christians were even allowed to worship in Mosques. And while the Christians in the Holy Land spoke Arabic or Greek, they sung many of the same hymns that were sung in Europe.

The Crusades originated for all the wrong reasons. Including especially their origin as a way to stop the murderous fighting between the Barons and their victims. The serfs or slaves. Christians killing Christians. It was getting so bad the serfs could not plant and harvest enough crops to feed people. And there was the very real danger of mass starvation. So Urban II and his spin doctors the priests got to work and fed their Christian victims with lies and falsehoods about the Muslims and their supposed atrocities against Christians in the Holy Land.

After a few years of this an army of perhaps 500,000 was assembled to march off to the Holy Land. Because of famine and minor skirmishes along the way maybe only 35,000 or so arrived in the Holy Land. For some time they were not successfully militarily against their victims. But finally through treachery involving a Christian commander of a Muslim force they succeeded. Odd the Muslims would have some Christians as commanders, yes? But maybe they really were tolerant?

Once inside the fortresses the Christians butchered everyone who was not a Crusader. Including all the Christians. All that they could find. Except in Jerusalem, where one small group of Christians survived the massacre. As Weaver says, "Huddled together, awaiting death, unable to speak the foreign tongue of the bloody killers [fellow Christians from Europe], they sang a mass in Greek: the Crusaders, recognizing the tune, swerved by and left them alive".[36]

Contrast this with the actions of the Muslim commander

---

36 Henry Grady Weaver, *The Mainspring of Human Progress* (The Foundation for Economic Education, Inc., New York, 1997), 109.

Salah al-Din, better known as Saladin, when in 1187 he recaptured Jerusalem. His forces fought only with the Christian knights. There was no butchery of civilians of any faith. And when the battle was over Saladin told the Christians they had from thirty to ninety days (varies in history) to gather their possessions and leave the city. There were few remaining Christians who were not Crusaders, but those that wanted to stay could of course do so.

There was some negotiation with Balian of Ibelin, the Lord of Ramla and Nablus (who had broken his word to Saladin) on ransom for individuals. In the end, most were allowed to go free.

Balian had asked Saladin for permission to retrieve his wife and family from Jerusalem before the siege. Saladin agreed on a one day visit. But then Balian decided to stay and rally the defenders.

In spite of this, Saladin arranged for an escort to accompany Balians's wife Maria and their children and all their household to Tripoli.

After the siege and fall of Jerusalem there was much grumbling among the Muslims and other non-Christians, that the Crusaders were robbing everybody and preparing to take the property with them. They appealed to Saladin to stop this. Who effectively said: "Hey, look, what can I do? If I tell them to stop this they will complain that Saladin has gone back on his word to let them take their possessions. There is not enough time to examine every item and have them prove it belongs to them. It is better to let them go, be rid of them. And have them perhaps bless Allah and Muslims for their tolerance and forgiveness". There was no thought of revenge against civilians, which was not the original idea of the Prophet Muhammad and not the way of Muslims at that time.

After the death of Rome the Saracen civilization that sprang from the creation of Islam created a bright light in part of the world. From the Ganges in the east to Spain in the west. And from North Africa to Venice, which in a sense was part of a "business empire" within many political entities.

Results of mysticism

Religion—mysticism—has provided no value to Homo sapiens. Well, has it? Many of the major wars have been religious wars. When not the reason for the war it is used to justify its "rightness". Each side feeling their god is on their side. And both sides may even have the same god. All a bit confusing…

Politics

This is the disaster that has haunted us since the first tribal chief took control of his victim's lives. And has continued for thousands of years, With the tribes becoming ever-larger and tribal chiefs ever-more megalomaniacal. Until you have a "Superpower" headed by a person who is "the most powerful person in the world".

Then in the name of "entertainment", film directors transport this destructive nonsense into space and the future. With "galactic federations" and "battle stars". Totally forgetting that those species having a societal system based on coercion can not survive long enough to acquire the technology for space travel. As Homo sapiens are proving.

Galambos always had a problem with the term "astronaut", one which I can understand. He said at best we had only "planetonauts". And hardly that, as no Homo sapiens have ever set foot on another planet. And the Russian "cosmonaut" is even more ludicrous. Implying travel beyond our own Bruno Galaxy.[37]

Politics would have us believe that Bush—*Dour Leader*—is "the one without whom we could not live" if he were suddenly not there. We would be bereft and powerless to run our

37 As Galambos pointed out, this would be the proper name for what was originally referred to as the "Milky Way". Something left over from the time we knew relatively much less about our world. And although Bruno was not the first with the thought of an infinite universe and worlds, he was the first prominent one after the invention of moveable-type printing.

own lives. At least that is what most people feel. Although perhaps hating the person at the same time. But is that really true? That we could not get along without a ruler?

The unfortunate (for him personally) Ariel Sharon had a stroke earlier this year (2006). Before that time many could not imagine like without him. Yet suddenly he was gone. Within a few days it was almost as if he had never been there. And in fact it made no difference in principle, because the *system* continued. Only under a different name.

In monarchies the chant is:

The King is Dead! Long Live the King!

Under a democracy the principle would be the same:

The Incumbent is Out! Long Live the Incumbent!

And it isn't even the incumbent. The "chosen one". Rather it is the *system*. A *system* that continues regardless of the individual politician entranced and entrapped by it. There could be an avatar, and the unthinking masses would still bow and obey.

If we want to continue with the system we could save a lot of money be deleting the human. Random laws could be generated by computer. And make as much sense—perhaps more—than those generated by unthinking politicians.

The thing that says this system is not good are the results. The *candidate* has brought with them all the societal problems that ever existed. Unsolved. And is adding more all the time. Reminds me of the film *The Manchurian Candidate*. Someone masquerading as a person they are not. The *candidate* is masquerading as someone who can help you and solve problems.

The candidate—the "political agent"—catering to those unwilling to assume the duties and responsibilities of individual thinking life.

How could the specie continue to ascend? It couldn't, and is in free fall to non-viability and then extinction. If we are not already non-viable we are very, very close to it.

The Political System

You must understand the reality of political systems. Of a

societal system based on coercion. And especially the popular variant known as "democracy".

Political System

A mechanism that controls the life and property of others.

Well, doesn't it? Isn't this the basic reality of the system. If you don't believe it, try saying "No, I will do as I want. Not what you want". And see what happens.

Although I do *not* recommend doing this.

Democracy

A political system wherein the controller is selected by majority vote of those controlled.

This differs only slightly from a *monarchy*:

Monarchy

A political system in which the controller is usually replaced upon death—natural or otherwise—by a family member.

And a *dictatorship* is different only in style:

Dictatorship

A political system wherein the controller is replaced by death or force.

You see how proper, absolute definitions and semantic precision can help to clear up things? To allow us to create new and useful knowledge and to solve problems? Democracies, dictatorships and monarchies only differ in style of implementation. Not in the reality that they are a society based on coercion. And that the true, realistic result will be the same. That societal problems will be created, maintained and increased rather then solved.

Think of what Darwin told us and what happens in a democracy. A political agent—incumbent or aspiring—listens to what individuals want. If they are successful in convincing their victims—usually referred to as *citizen*—they can provide what is wanted they will be elected or not become unemployed.

They will create unnatural laws that attempt to favor a majority—but not all—of their victims. Recognizing they can not favor all. Those who are not favored will find they can not do what they wish with their life. There will be required behavior and things that may not be done. Under penalty of death. If this were not so, political states would collapse. Read "Cat License or Death" in my book *Cowardly New World* or at <http://www.aristarchus.org/Cat-License-or-Death.php>.

Think about the result of this system. Two basic things. One is that individual life and freedom is not possible. In a political system there will always be laws prohibiting or requiring things. Secondly, that this must be an ever-increasing process. If the incumbent does not pass new laws why are they needed? They will soon become unemployed. Or a fate worse then death for a politician: unpopular and irrelevant. Poll ratings are everything.

This results in *Government Without End.* Something I suddenly realized in the summer of 1956. Under democracy eventually everything would either be required or prohibited. No need to think, just obey. The brain slowly dies. The body dies. And the specie dies. Think about it. Isn't that what is happening?

Government Without End

Background

The thought occurred to me in the summer of 1956 when I was living and working in Japan. I no longer remember what brought it on. This was a time of relatively much greater individual freedom in the US and the world, and far less political interference in individual life. In retrospect, an almost idyllic time. But I suppose I must have had some altercation with the regime, even though I was on the other side of the world.

But I suddenly realized that by the very nature of a political society—a society based on coercion (and fraud)—that laws must increase on a continuous basis. After all, if they did not do so, what was the purpose of the political agent? That unnatural laws must proliferate until everything has either been required or prohibited. There is no longer any need to think. Thinking life would cease to exist. Except perhaps for Big Brother. But it would seem to be a lonely world for him.

Basis of Politics

The basis seems to be that most people cannot think properly. Can not properly perform the duties and obligations of individual thinking life. Perhaps even that most are "evil". And will inevitably do things that injure their fellow thinking beings. So thinking and action must be controlled. To prevent chaos and disorder and provide security and tranquility in society.

And if you look around you, you must agree this system works splendidly, yes? It must, as I don't see anyone making any serious effort to change the system. Only a little tinkering here and there. There is always dancing in the streets as *l'ancien régime* is out and *le nouveau régime* is in.

Politics in Action

In a democracy political laws will be continually created. Until there is nothing left to forbid or require. Why is this so? It is required by the system. Think about it.

What must the new "electee" do to ensure they are successful in office? With "successful" meaning they do not become unemployed.

To be successful in a political sense the electee must find something that needs "regulating". Something that must be forbidden or required. Something that will satisfy a potential "need" or desire of a majority of electors. The politician—the ruler and ruling party faction—must do something to restrict the freedom of their victims. Usually referred to as "citizen". Although the British use of the word "subject" is nearer to reality.

Under politics the satisfaction of one individual need always represents a claim on the property of one or more other individuals. If not, the electee and their access to political police powers are not needed. Given the diversity of individual hopes and dreams does it not seem preposterous that one person can or should represents millions of people? All with differing hopes and dreams.

But the restriction of individual freedom is thought necessary to prevent chaos in society. After that statement I should not really need to say more, true? Let the results of whether or not politics works speak for itself.

For one individual to assume they can do this—and others to believe it is possible—beggars the imagination of a rationally thinking individual. It would be a bit like a husband and wife retaining the same attorney in their divorce. There would be conflicts of interest and other realities that would make the outcome unacceptable to both parties. Why do we tolerate this in our societal system?

Each electee has an implied contractual relationship with their individual electors. One that violates all rational and legal contractual concepts. When the electee fails to perform according to the contact with each individual elector, the electee can not be fired. And they will retain their pay and sumptuous offices, travel opportunities and many other benefits. Maybe even freedom from traffic problems when they travel on business. As they control the police and associated traffic control monopolies.

They determine the needs of their constituents by listening to them, Reading newspapers, listening to the TV news. Perhaps reading their mail. Most likely when they were running for office they already had a list of things that needed doing.

The actual, true needs of a thinking being are small. They need water and food to survive. And shelter in many climates. Yet beyond those basics, human needs are virtually unlimited. The basic characteristic of wanting to acquire the most amount of property for the least amount of effort.

With variations consistent with the uniqueness of individual life. Unique and different hopes and dreams for the future of the individual. And differing abilities to achieve them. Further, this is a dynamic situation, constantly changing throughout life. The permutations are very large. And attempts to regulate these characteristics—the function of politics—adds additional unpredictability. As individuals modify the actions associated with their hopes and dreams to cope with the regulation. This creates what mathematicians refer to as a *chaos system*. As you may notice from history and the daily news?

As stated elsewhere a basic law of nature is that individual thinking beings seek to acquire the most amount of property for the least amount of effort. Were it not so, thinking life would not evolve. If it appears a person might receive something for a very low cost to themselves or even "free," then of course it is "needed". And if experience shows that others have done this or similar things, then others will want the same. Or more. This is ready made for the electee.

They must quickly start working to create "bills, measures" and other things that will become unnatural laws. Political laws. Else they will have no accomplishments to point to. And there is always another follower who will be waiting in the wings. Politicians can not really be leaders. They must be followers. Following the wish lists of the electors. As confirmed by their devotion to polls and their results.

If they are not followers, then they are dictators. Such as George W. Bush. He is trying to rule by decree without actually dissolving the Congress and Senate. But give him time. He has said that it might be necessary to postpone elections if there is a large enough terrorist act. And it not unknown for politicians to create such events. Such as was done to start the American War in Vietnam. And also the one in Iraq…? And there it was only the *suspicion* that something *might* happen.

But generally they are followers, and the more successfully they do this the less chance of joining the ranks of the unemployed. If they try to be a leader and lead people in a direction they do not want to go the politician may join the ranks of the unemployed. Much better to be a follower.

This creates a system of ever-increasing needs. From seeing the example of others who have had laws created in their favor or the need to create an opposing law. Further, as individuals are denied the right to think for themselves, making mistakes and learning from them, they become less competent. And need ever-more things done to protect them from the world around them.

This is why people become unemployed. They do not understand the true, proper, realistic concept of job security. And do not create it for themselves.

Laws become ever-more complex with ever-more side effects and collateral damage. The "law of unintended consequences" is very obvious in politics. Unnatural laws are never cancelled. New ones are simply added to modify existing ones. Which is why ignorance of the law *is* an excuse for not knowing the law. This is confirmed by the highest court in the land. Which has an odd number of members so there can never be a tie. And many decision are reached by a majority of one.

The only way this process of *Government Without End* can be stopped is by replacing the societal system that allows, requires and creates this. History shows us modifications and alleged "improvements" to the current system or "reform" accomplish nothing. The system is based on a flawed premise: that it is moral and rational to control the lives and property of others. That some if not all thinking beings require this.

This is not true. It has never worked, and of course never can work. But those laws of nature that explain how individual thinking beings respond to outside stimuli can be used to create a rational and durable societal structure. One that does work.

Results

Any rational person would wonder from where it is we receive value in life. Why we have a societal system that does not allow us to solve any of our societal problems. Is this—the daily news—the best that we can expect in life? Remembering that it seems to get worse rather than better.

We have ceded responsibility for the duties and obligations of individual thinking life to others. We no longer have control over our own lives because we have said we do not want this. And created a system that, rather than taking care of our lives for us, has made a complete mess of things. And we do not know how to do anything about it. Or worse, we no longer care about the results. Except to complain and blame everyone except ourselves for allowing this to happen. And then go shopping, to a sports event, or on holiday and try to forget the problems.

Chapter 5

The Natural Society

Synopsis

We have a societal system based on coercion. Which has prevented us from solving even one of our societal problems. Rather, they become ever-more endemic and intractable.

A person must wonder why we do this. When the alternative is so simple and obvious. Create a societal system based on the nature of man the specie.

Background

The title is with credit to Andrew J. Galambos.[38] It describes a societal system with a few simple rules that are consistent with the laws of nature. No more than three rules. Which are in fact laws of nature. A societal system that will therefore work.

I did not plan to write this. I planned to leave you hanging at the end of the previous chapter. To let you be angry with me. Or, better, to require you to think. For it is only through thinking that we can disabuse ourselves of our many societal fallacies. Those related to a societal system that obviously has not worked. Further, it should be obvious that it can *never* work. And to think about one that does work.

If you think on your own, in a vacuum so to speak, it could take too much time to come up with a workable alternative to our current societal system based on coercion. After all, it took me most of thirty years to understand it. Even though the principles were laid down for me by Galambos. And be able to explain it to you. As you may notice, I am still a bit deficient in that area. So I will mostly write about the principles.

---

38 Andrew J. Galambos (1924-1997). Creator of the Science of Volition and of the Laws of Volition. Integrator of the Volitional Sciences.

The more you can think on your own and the more resources you expend in doing this, the more you will value the solution. And understand it. As Galambos would say, you will have a "proprietary interest" in the solution. By virtue of your time investment in it.

Although Galambos may not have been original on this, he did describe what I think is very good idea in teaching. The teacher will explain the problem. And start to explain possible solutions. But not go all the way on each path. Perhaps starting many times and having some false leads. But sprinkling the principles around.

With the result that most if not all of the students can eventually deduce the solution before the teacher states it. When this happens, learning occurs. When the student creates the solution in their own mind. Using a principle they may use to solve other, similar problems.

Please do not think I am speaking down to you by my polemic way of saying things. I wish and intend to speak up to you as a fellow thinking being. The value of whose life is huge, almost immeasurable. How do you value a human life? Well, actually, there is a way…

The value of a human life lies in the new and useful knowledge and other products created by the individual. Thus, the value of the life of an Archimedes, Newton, Einstein or Galambos is very great. And can actually increase even after they are dead. As with the value of the individual members of the Beatles. The ballerina Margot Fonteyn, Frank Sinatra, Barry Manilow, Barbra Streisand, Britney Spears. The individual who removes the trash from your house and street. As all of these have created value in society.

The value of the life of a Roosevelt (Theodore or Franklin), Kennedy, Johnson or George W. Bush is zero. Actually not zero, but negative. What have they provided? War and terrorism. Failing education, healthcare and pension systems. The huge budget deficits that mortgage the future and turn American government bonds into junk bonds. Crime, hunger, famine. The largest per-capita prison population in the world.

I suspect you can list more of the products of he and his system.

More than that, politicians control the life and property of every single thinking being on the planet.

There is a simple test to determine if an individual is or was important in life. Has importance for you as an individual and for the specie.

As yourself a question: how would your life and the lives of others be or have been without this person? The same? Worse? Better? Without your trash man or favorite entertainer it would not have been as good, yes? Without Alexander of Macedon or Napoleon the lives of millions of people would have been almost immeasurably better.

Why do many consider Alexander and Napoleon to have been important? Perhaps there is something wrong with our system of value? Maybe with our understanding of where and from whom we receive value in life?

What about those who have created the recent, great advances in biological knowledge? Contrast their lives and importance with that of the politicians that control the healthcare delivery system. Yet please think of the fact that it is the *system* that is bad, and not so much the individuals.

Think of George W Bush, and Steve Jobs of Apple. Which one has created property and things of value? And which has destroyed vast amounts of property and caused the death of hundreds of thousands of people?

Before his terrorist activities in Afghanistan and Iraq are over *Dour Leader* and the *Axis Powers* he heads—the *Gang of Four*—will have been directly responsible for the death of up to a million people.[39] For which he will receive a splendid pension and honors from the American political state. Is a societal system that allows—even requires this—something of value for a thinking specie?

To verify that his acts constitute terrorism you need only

---

39 The Washingon-London axis, and Aznar, Bush, Blair and Howard. The original *Gang of Four*.

understand and use a proper definition of terrorism. Do you not you believe that most of those hundreds of thousands who have died felt terror before and during the time they were dying? Isn't dying and even the thought of dying a terrifying experience? And that their families and others felt terror as they witnessed the act that killed family and friends?

Would it not be better to understand the simple cause for the initial acts that caused his terrorist activities? And to fix those so there is justice and not more such activities? After all, it is very simple to do.

This discussion of the subject of importance can not be a complete and exhaustive discussion. There is a discussion of that subject in my earlier book *Cowardly New Word.* And will be a much more complete discussion in the forthcoming *The Conquest of Ignorance and Human Survival.* You can read my essay on importance at:
<www.aristarchus.org/importance-in-history.php>

If I can not say something that will lead you and us to survival as a specie then I have failed. So I do have some incentive. More, actually. As though I am old and at the end of my life, I am incredibly fortunate in having a wonderful young wife. More amazing, a five year old boy and a 2 ½ year old girl. I must try to fix things so they will grow up into a world with less and less disorder. Which will occur with more and more individual freedom. Culminating in a societal system for Homo sapiens that actually works.

Along the way taming the chaos that will result from Global Warming, and then solve that problem also. And fulfill our cosmic destiny of exploring and colonizing the universe. A moral form of colonization, more fully described in *The Conquest of Ignorance and Human Survival.* Where I also explain why we need have no fear of intelligent, extraterrestrial beings. Rather, we should hope they visit soon, as they would have or could quickly outline the solution to our climate change problems.

Historical Background

General

Moses was seemingly the first individual to "write" about individual freedom. The first commandment suggests rejecting the multiplicity of the gods of the time and having one god only. This is a step forward toward individualism. The sixth commandment stresses the sanctity of human life, also an expression of individualism. The ninth stresses the right of free speech. Clearly an individual act and right.

Abraham and the Jews appeared on the scene about 4,000 years ago?[40] And with the Jews and Abraham, Gideon and Samuel a concept of individualism began to rise. After Gideon had freed the Jews from the Midianites the people asked Gideon to become their ruler. But Gideon said "I will not rule over you, neither shall my son rule over you: The Lord shall rule over you". He did not favor the idea of one mortal man to rule over others. They should have individual freedom under their belief in a single god.

A hundred years later the Jews were appealing to Samuel to be their ruler or to provide them with one. As now, with people pleading, "Oh please, rule me. Take control of my life and property. The responsibilities of individual thinking life are too much for me". But Samuel would not do it. And described what would happen to the Jews if they had a temporal ruler. Describing all the onerous things a political state does to the individual. What he said was:

> *"This will be the manner of the king that shall reign over you. He will take your sons and appoint them for himself, for his chariot and to be his horsemen; and some shall run before his chariots. And he will appoint them captains over thousands, and captains over fifties; and will set them to ear his ground, and to reap his harvest, and to make his instruments of war.*
>
> *"And he will take your fields and your vineyards, even the best of them, and give them to his servants. And he will take the tenth of your seed, and of your vineyards… And he will take your men ser-*

---

[40] Please, do not worry too much about dates. I am talking principles.

vants, and your maid servants, and your goodliest young men… and put them to his work… And ye shall cry out in that day because of your king which ye shall have chosen you."

But the Israelites were not convinced and said "Nay, but we will have a king over us. That we may be like all the nations, and that our king may… go out before us, and fight our battles". So they would not need to think and worry about life. Just be told what to do and then complain about it. And eventually they did have their way.

They had a string of kings. Including Solomon. Who with increasing taxes built magnificent palaces for himself. As rulers are prone to do. They got their nations and then as often happens their ruler was defeated in battle. And they became part of another nation and slaves of yet another ruler.

Most of the old world rulers hated the Jews, with the word *Jew* symbolizing freedom and individual initiative. Even when they had long ceased to have it.

The next appearance of some sort of individual freedom was with the Greeks. Although there was a class structure and relatively less freedom for the lower classes. Again the same as now. They had implicit faith that their government would be good to them and take care of them. Now, more than 2,000 years later there is less faith in this. More a resigned acceptance that "this is the way it must be. Our ruler, right or wrong".

As now, the Greeks and all others labored under the delusion that their system was the way to "peace and plenty". Not realizing that unrestrained majority rule not only destroys freedom but brings with it "the rulership of mediocrity".

The beginning of the end for Greece was with the glittering Pericles. Who maintained his dictatorship through lavish public spending. Almost 3,000 years and we still have not learned. Through attempting to provide everything for the people who "needed" and wanted something. Then because this was not working and he needed a diversion, he went to war with Sparta. Another tactic of politicians to divert attention from internal unrest. Problem is, he lost the war.

Greece did produce a number of major individual, independent thinkers. Archimedes, Aristarchus of Samos, Hippocrates, Hero of Alexandria. It is Seekers such as these that created a foundation for science.[41] One that when Greece disappeared as a political entity, the Saracens would preserve and add to. And eventually set the stage for another explosion of knowledge that would create the Industrial Revolution.

Two thousand years after Abraham, Christ appeared on the scene, ushering in the Christian era. Because a societal system based on coercion is so debilitating for the individuals of a thinking specie, hope springs eternal that something will happen to deliver them from this system promoting slavery and non-thinking. But they received only a false refuge. A step backwards from the words and ideas of Moses into a chaotic system of retribution and immorality. With everyone begetting people without benefit of marriage and endless retribution. Casting people out, having them roast in hell and other unappetizing endings.

As an individual Christ was quite possibly a good man. We can not really know because of the conflicting fables surrounding his life. Created over the centuries after his death. He had no biographer during his lifetime or the early years after his execution by the Romans. For trying to compete with and supplant their system of gods. Something that would destroy the fabric and basis of the state and attack the divinity of the emperor. Any political ruler would consider this to be treason. And arrest and execute the person promoting such traitorous, treasonous and treacherous acts. This is a political necessity and "right".

There was no biographer of Jesus at the time for the simple reason he was not well known. And during and after his life there were competing prophets that were as well or between known than Jesus.

---

41 With credit to Daniel J. Boorstin for the concept of the Seeker from his book *The Seekers*. (Vintage Books, New York, 1999).

The Roman civilization into which Christ was born created about a seven hundred year hiatus for the concept that the individual should remain in control of their own life. The only thing they created of tangible value were their monuments, sculptures, roads, buildings and aqueducts. As I have said, their creation of laws was largely a disaster for Homo sapiens. Codifying the concept of a societal system based on coercion. You should realize this. It is the basis for and the cause of the chaos in political societies. Whether they are democracies, dictatorships or monarchies.

The thing that would begin to change this was the Arab and Muslim civilization of the Saracens.

Saracen Civilization

Most people are not familiar with the Saracen civilization. They were predominantly Arabs and Muslims. *Saracen* was the name given to them by Europeans who did not understand them, were afraid of them and hated them. Perhaps secretly envious of them. Because during the Dark Ages of Europe the Saracens had a bright and shinning civilization.

The Christian Crusaders invaded and devastated a "modern" world beyond their ability to understand it. As Weaver said, "… the Crusaders cane into rooms that were like jewels—with tile floors and mosaic ceilings. Faintly scented air came in through the delicate latticework windows which permitted one to look out, though no one could look in".[42] He contrasted this with the European castles and their crude and rude stone walls and floors littered with rotting seeds and bones.

That "… the invaders came upon cups they mistook for rare jewels and which, in their excited grips, they crushed to splinters that cut their hand".[43] They had never seen the fine, sheer glass of the Saracens. Originally created by Syrian

42 Henry Grady Weaver, *The Mainspring of Human Progress*. (The Foundation for Economic Education, New York, 1997), 111.

43 Ibid.

craftsmen from the Sidon-Babylon area. Around this time the Venetians—part of the Saracen business empire—assumed the role of the glassmaking center of the western world.

They had scientific farming, producing a wide range of fruits and vegetables never seen before. The Europeans used wooden plows that barely scratched the surface. But from the Ganges to the Atlantic, Saracen farmers were deep plowing and contour plowing. Irrigating and rotating their crops. Fertilizing them. And with the vast commercial network—almost like a common market—they were shipping their products thousands of kilometers.

A Saracen house might have running water, and a form of air conditioning. They had fine silks. Their chain mail would fall into a ball in your hand. Weigh very little, yet stop an arrow. While the armor of the Christian knight often equaled their body weight. The knight probably took at most two baths in his lifetime. While the Saracens bathed five times a day. Their women had a wide range of cosmetics and perfumes available.

Their world was one of great religious tolerance, quality of life and knowledge. They collected and translated the work of the Greeks into Arabic. Then their own agriculturalists, astronomers, geographers, mathematicians, surgeons and others greatly added to that knowledge.

The Saracens actually had a "business empire" inside a multitude of political empires. It was business that tied them together. They used double-entry bookkeeping. Had a postal system that could take something 2,000 miles (3,200 kms) in ten days. Providing a receipt within that time.

They flourished from about 800-1500. Their business empire running from the Ganges to the Atlantic. And from North Africa to Venice. In the 1200s the Venetian surgeons had a rather complete knowledge of anatomy, gained from the Saracen texts that had been translated into Latin. They were performing surgical operations and dissections at a time when the Inquisition was burning people at the stake for the same thing.

It was the translation into Latin of the Arabic texts in the 1200s that was the proximate cause of the *Renaissance.* And perhaps "godfather" to the *Industrial Revolution.* For without these texts and the knowledge they contained Galileo and Newton could not have done their work. And Columbus and the Portuguese navigators would not have made their "voyages of discovery". The Europeans learned their cartography and celestial navigation from the Arabs. The word *azimuth* is from the plural of "as-samt", meaning "the directions". And of course the English word *algebra* comes from the Arabic "al-jabr" meaning "the reunion of broken parts".

The Saracen civilization died in the late 1400s. Murdered by Christian politicians. Until the time of the Christian victory over the Moors (Arabs), Moorish Spain had been a place of religious tolerance. Christians, Muslims and Jews worshiped freely, side by side. The same system that had existed in the Holy Land until the Jihad by the Crusaders.

After the Christians defeated the Moors they launched a "Jihad" against Jews and Muslims. Murdering many without a word. The rest told to convert, flee the country or die.

You can read more about the ancient Saracen civilization in *The Mainspring of Human Progress,* by Henry Grady Weaver.[44] It also has a bibliography.

American experiment

Weaver refers to the time of 1650-1850 in the American colonies as "the third great experiment with individual freedom". Which tragically ended with the establishment of political institutions on the West Coast beginning in the 1840s. Individuals were in a trap with nowhere to go. You could not "move further west" and be in charge of your own life.

It would still be perhaps another fifty years, until about 1900, that the politicians really consolidated their control over individual life. Stamped out individual freedom. The members of the "House of Representatives" and the "Senate" de-

44 See the bibliography for more details.

cided to work together for the benefit of their members. And the detriment of their victims. Not unlike the Council of Nicaea in 325 when the civil and religious branches of politics decided to work together.

Just today (August 16th, 2006) I received Boorstin's *Hidden History* in the post.[45] And almost immediately found a quotation from Jonathan Swift. Who said:

> *"Government without the consent of the governed ... the very definition of slavery."*[46]

Sadly, Americans believe the myth that they are governed with their consent. If you have ever tried to revoke your consent you will know the true condition is as Swift said: slavery. Or if you try to change the rules under which you are ruled. In a sense each individual should have their terms and conditions under which they are ruled put into a contract. Which of course must include their hopes and dreams for their future. With the notice and clause that this may change from time to time based on the experience of life. And have their political agent agree to and sign the protocol. But then why not just take control of your own life? It is simpler.

Each of the 21 million or so Americans in California that are ruled by a Senator (two senators for the 43 million of so victims) should have said individual agree to and sign their terms and conditions. This would keep them busy, just making laws that favored you. While somehow not adversely effecting the life of another person, yes? Do you think this can be done? The objective is rational and moral, but its implementation is not possible. And why bother? The natural condition is that you and every other individual thinking being can do whatever you wish as long as you do not *intentionally* create an adverse effect on the life of others. The agreement, the law, the Senator, and the concept is not needed. Espe-

---

45 Daniel J. Boorstin (1914-2004). American historian and writer, Librarian of Congress (1975-1987). Pulitzer Prize winner for his book *The Americans: The Democratic Experience.*

46 Daniel J. Boorstin, *Hidden History.* (Vintage Books, New York, 1989), 66.

cially because—as history shows us—it does not work. And as thinking about it tells us, it is immoral.

Basic Concepts

General

What is it that tells us what did not work and what will work? And who provided us with this information?

Scipio

If anyone thinks of Publius Cornelius Scipio (235-183 bce) they almost certainly do not think of him as a person who promoted individual freedom. After all, he was a Roman general. And, yes, he did save Rome from conquest by the Carthaginians. Which in turn ensured Europeans and thus Americans would have an Occidental and Christian heritage rather than Oriental and Muslim. But a promoter of individual freedom?

He did this by his simple, eloquent and elegant statement when offered the job as sole Consul (dictator) after his victory over Hannibal on October 19, 202 bce. At about 6:00 pm, Roman Central Time.

He responded by rejecting the offer, with the explanation that "No man should rule other men. And certainly not for a lifetime". To accept would have made him the ruler of the world's first "Superpower". Somehow he knew this was not right. Whether it was from the standpoint of being moral or not moral—or simply irrational—we can not know for sure. I suspect he understood both concepts. This from a man still only thirty-five years old. In the prime of life.

Can you imagine *Dour Leader* turning down such an offer? After being accorded his *Triumph* for defeating the Iraqis. As he will surely do. After all, the US has a population ten times that of Iraq. And I think a GNP of 10,000 times that of Iraq. And well-acknowledged that the US has the most formidable military machine on the planet. Having successfully invaded Grenada, Panama and other countries.

From history we of course know that Scipio was correct in his statement. As the concept did not work. And the Roman empire soon (in historical terms) collapsed from a societal system based on coercion, with an extreme case of socialism. The same thing killing our modern political states. And the specie.

Scipio did eventually agree to accept for the usual one year term. I suspect the main reason was to honor those who served under his command while protecting his beloved Rome.

Paine

The early life of Thomas Paine (1737-1809) gave few clues to his eventual greatness. Unlike most major innovators he did not do his major work until relatively late in life. It was nineteen days before his 39th birthday when *Common Sense* first appeared. Although Benjamin Franklin must have sensed something earlier. When he suggested Paine go to America. And even provided letters of introduction for him.[47] We thus owe yet another thing to Franklin. An already remarkable man.

Prior to the publication of *Common Sense* no one in the Colonies thought of independence. They only wanted the king to be a better, nicer king. After January 10th, 1776 almost the only thing people thought of was the need for independence. Which in principle is individual freedom. Freedom from the tyranny of what was still the Age of Serfdom in Europe and rest of the world.

Although I can not really discuss it here, the founding of America was the third and so far last attempt to create a societal system based on freedom. You can read about this in *The Mainspring of Human Progress*, by Henry Grady Weaver.[48] In

---

47 A letter of introduction is the origin of the concept of the passport. Which has been completely perverted by modern politicians. Impudently and insolently assuming they have control over your travel.

48 You may learn more about this book by visiting the bibliography in the appendix.

which he also mentions some other good books on the subject.

Paine was a very clear thinker. And he wrote clearly and concisely of what he thought. This is evident in *Common Sense*, *The Crises*, *The Age of Reason*. And in the Declaration of Independence, of which he is the author. *The Age of Reason* earning him the enmity of non-thinkers worldwide. Who said he was an atheist. Yet it is very clear that only those who had not read the book or not understood it could say such a thing.

For on the first page of *The Age of Reason* he said "I believe in one God, and no more; and I hope for happiness beyond this life". Does that sound like the thinking or words of an atheist? He merely had some very rational questions about some of the things said in the Christian bible.

What I state here are not really "laws of nature" in the common understanding of that concept. They might more correctly be termed "observations on human nature". Although individual thinking beings must follow the laws of nature. Such as Galambos' Laws or the *Universal Laws of Volitional Behavior*, which will be discussed later in the book.

One of the things Paine said in the Declaration of Independence is: "We hold these truths to be self-evident, that all men are created equal…" His intended meaning is that of other natural philosophers of his time. That Homo sapiens were created *equally* under the laws of nature. But that in accordance with the as-yet unarticulated laws of Darwin they were not actually *equal* to each other.

Paine also said "…that among these are Life, Liberty and the pursuit of Happiness." This anticipated Galambos' Laws, specifically the Second Law of Volitional Behavior. Which states:

Rationally thinking beings seek happiness.

Although this will be covered later, this is extremely important. For it is the regulatory mechanism in a rational and durable society that replaces our current concept of "regulating" individual behavior by coercive means. Unnatural laws and a monopoly on police powers to enforce them.

These were all things anticipated by Paine. And things he probably would have elucidated had he not been distracted first by the French Rebellion.[49] Then later by bad health as a result of his imprisonment in France.

The last thing mentioned in the *Declaration* appropriate to this discussion is the statement:

> *"That whenever any Form of Government becomes destructive of these ends, it is the Right of the People to alter or to abolish it, and to institute new Government, laying its foundation on such principles and organizing its powers in such form, as to them shall seem most likely to effect their Safety and Happiness."*

Paine had not yet refined or properly defined the word "government". This was left for Galambos. Although I think it is clear from the writings of Paine that he did not intend the word to mean "political régime". In the sense of those in the past that ruled through coercion. It is a bit of unfinished work on the part of Paine that he did not clarify this.

What Paine was saying was that if "government" did not create liberty and the ability of the individual to pursue happiness as they defined it for themselves *as individuals*, then the job was unfinished. Which of course it is. As under the American style of democracy there is and can be no "liberty". Certainly not in the sense of complete individual freedom. Based on the concept that each individual thinking being is free to do any and all of those things that do not *intentionally* interfere in the property of another individual. That each individual is totally responsible for the moral and rational discharge of the responsibilities associated with individual thinking life.

This is the essence of individual thinking life. Wherein individuals can follow their own individual hopes and dreams for happiness in their life. Which by definition does not inter-

---

49 This event is generally referred to incorrectly as the French *Revolution*. Although it was clearly a rebellion against existing authority. And replacement of that existing authority by a different political régime. This is not new, which is the proper concept of "revolution." Which means a "turning around," or something new.

fere in this same activity for other individuals. And being responsible for their own lives increases competence.

Since the summation of individual life is societal life and a specie-wide thing it also maintains and increases the quality of the gene pool. Ensuring survival, even prospering of the specie.

I am sure from things Paine said that he had read Thucydides. No doubt in *History of the Peloponnesian War*.[50] The translation of Thomas Hobbes (1588-1679) and also the autobiography (1671) of Hobbes. And that he was thus aware of the problems of democracy. Specifically Hobbes' statement that "... how silly is democracy, and how much wiser a single man is than a multitude". With "multitude" referring to what we often call the "silent majority". Which is never really silent. Or the term the Communists used: the *proletariat*.

Some of this is a bit odd as Hobbes himself believed in a strong monarchy. Albeit based on some nebulous "social contract" with those ruled. He also believed in determinism, rather than the concept that a person was in control of their own life.

For us as a specie in attempting to ensure that we as a specie will be able to handle the forthcoming problems related to Global Warming, we need to think of the following statement from Paine:

> *"These are the times that try men's souls. The summer soldier and the sunshine patriot will, in this crisis, shrink from the service of his country; but he that stands it now, deserves the love and thanks of man and woman. Tyranny, like hell, is not easily conquered; yet we have this consolation with us, that the harder the conflict the more glorious the triumph."*[51]

The time in the next ten years will test us to the limit. Mainly our ability to discard fallacies and to think rationally. To engage in *discovery*, as described by Szent-Gyürgyi von Nagryapolt. We can not afford to be "summer soldiers" or

---

50 Thomas Hobbes (1588-1679), *History of the Peloponnesian War* (1628).

51 *The Crisis*, by Thomas Paine. December 23, 1776

"sunshine patriots". But survival of the specie would certainly be a "glorious triumph", yes?

If we continue to follow *Bush Thought* and to ignore the hidden message in it, in twenty to fifty years at most we will run out of time. The disorder and *uncontrollable* chaos—not the *controlled* chaos we have now—will engulf all thinking beings. There will be no place that is safe on this planet.

Imagine what will happen when London, Hamburg, Bremerhaven, Amsterdam, San Diego, Los Angeles, Boston, Washington DC, Alexandria VA, Miami, and many of the Gulf Coast cities of the US are under water.

Darwin

I just now (September, 2006) noticed that Charles Darwin (1809-1882) was born in the year Paine died. They were successors in more than one way. Darwin spoke of the continuing evolution of the specie. Essentially a process of trial and error. Seeing what worked in the environment and what did not work. Continuing to fine-tune for the best result. For survival. For living things not able to make decisions about their environment this process can perhaps be optimized easier and quicker. But for a thinking specie that may change its environment—or more importantly, itself—it becomes more complicated. Even to the point where the specie may "interfere" in the process and derail it. When the specie alters itself. Or produces large and short-term changes to their environment.

Darwin tells us some very interesting things about the way species evolve. Some of which we have ignored when constructing our societal system.

One thing he tells us is that we evolve slowly. Continually trying to optimize ourselves. Both within ourselves and within our environment. The environment has generally tended to change very slowly, giving a lot of time for this process. Yet as we also know from nature, the environment can change quickly. And when it does, if the specie cannot adapt quickly enough it may perish.

We also know that within each specie, at least for the more complex ones, individual members of the specie tend to be unique and different from each other. This is a fact of nature.

In nature this is what is meant by "equality". By "being equal". Being "equal" under the laws of nature that say as individuals we will all be unique and different from each other. This produces "balance" in nature. Giving the specie a chance to survive.

Think if everyone were exactly alike. Something might happen in the environment that mitigated against survival of any specie with those "specifications" and they would be doomed.

Or if we suddenly say the fact that our "unique and different" characteristics are not acceptable. We seek to remake ourselves to become truly equal to each other. But that is not the way of nature and our environment has not changed. We no longer match our environment and its natural laws. This will produce disorder and chaos in our environment.

Further, there are still those individuals who insist they will be unique and different and "fight back" against a *societal* environment that is now different. This will produce additional disorder and chaos in the environment.

For thinking beings there are two environments. Or two components to our environment. The physiological and natural one, and the societal one. The former the one that gives the name to the specie. And the characteristics that makes us the dominant and most successful specie on the planet. The one that can change their environment if they wish and when needed.

The characteristics of a thinking being are that they think and take actions based on their thoughts. The idea being that if we make the "right" decisions we will survive and prosper. Both as individuals and as a specie. With "right" meaning the action is consistent with the natural world. That it does not cause problems in and with our *natural* environment. Which includes ourselves. If we cause disorder in this natural environment—by trying to flout the law of nature—it is not good for us as individuals. And thus not good for the society of thinking beings. What produces disorder?

The other environment is the societal one. Where individual thinking beings interact with other thinking beings. Our current societal structure is based on coercion. So the vast majority of interactions involve coercion. Interference in the property of others. If not actually interfering, talking about interfering. Because societies based on coercion create a world of "nitpickers". Those who do not like what someone is doing or how they are doing it. Think they have a better idea and want to impose it on others. Have unnatural laws created requiring or prohibiting something. To be backed up by the police powers of the political state to enforce the behavior.

This then has an adverse effect on the life of another thinking being. And this may invite retaliation. And unpredictable behavior from the effected individual and perhaps others. Both of these create disorder in society.

You really need to think about the "equality" thing. If we were all equal it would be a very dull world. In fact, if we were all equal we could not survive as a specie. We must have mathematicians, metallurgists, polymer chemists, radish and garlic growers. For our enjoyment we must have violinists, drummers, ballerinas, trumpet players and jazz composers.

This may seem like a tremendous digression. But remember: *all knowledge is related.* We must all strive to make a profit.[52] Because profit is *the indicator of success.* As Galambos said, that is its most important characteristic. No profit, what you are doing is a failure. Because profit can also be viewed as "money left over after expenses". If as a worker you do not have any money left over at the end of the month you may be operating at a loss. You can not survive this way. Neither can a business. But you are also a business. Your business is being

---

52 Remember Galambos' definition of *profit*: "Any moral increase in happiness".

successful at what you do in life. At selling to your employer the tasks you perform.

Your "business product" is the services you sell in the market. Which might be as a coal miner or violinist. If you do not make a profit you need to either improve your product or perhaps start to produce a different product. Maybe you can make more money—and have a more enjoyable life—as a violinist than a coal miner. Which means you must adapt to the market. Enough of that diversion. But please think about this as part of surviving as a viable member of the specie.

Darwin spoke of "survival of the fittest".[53] In terms of a thinking being what might this mean? Maybe it is connected to or equated with "competence"? If the speed of a cheetah was suddenly not enough to catch its prey would it survive? If a thinking being can not think well enough to survive in its environment it also may not survive. As I have said, the ability to *think* is what has given as the edge as a specie. For we cannot run as fast as some animals. Relatively as fast as some insects. We can not fly like the birds. We don't have the absolute or relative strength of many animals and insects. We can not swim like the fish.

But we can *think*. And build airplanes with which we can fly as the birds. Design devices that allow us to survive in the realm of the fish. With a single crane we can life more than billions of insects could lift. In fact, we are able to think so well that we can no something that no other specie on the planet can do: We have developed that ability to destroy our own specie. And many other species.

And so far we have proven that we are unable to design a societal system that can prevent that.

The uniqueness of individuals

Each and every example of Homo sapiens is unique and different from all others. Both physically and "intellectually".

53 Although he did not use this phrase in any of his work, he expressed admiration for it. This phrase was coined years later by Herbert Spencer in *Principles of Biology* (1864).

In their "primary property". The intangible result of their life. What is critical to the failure of politics as a societal system is the individual hopes and dreams for individual life. The ability to implement and realize them. Further, this is a dynamic process. Constantly changing throughout life.

Political result of this uniqueness

The societal system we have adopted for ourselves says that these unique hopes and dreams must be controlled. Limited. In politics "regulated".

What happens when you prevent a person from realizing their hopes and dreams? Inevitably there will be depression, frustration, irritation. And a search for another way to do things. The search for a "loophole". And of course the function of the politicians is to close the loophole. Then maybe another attempt at the dream. Or maybe not. But soon it stops. Although the depression and irritation will not.

With millions of people there will be perhaps tens of millions of hopes and dreams. Then variations as the search for loopholes goes on. The permutations—possibilities—can become very large. Unpredictable actions. This is a major cause—perhaps *the* major cause—for disorder and chaos in society. Because you then also have the marching and protesting to go with this. Even the bombings, as has happened to abortion clinics.

But worse can happen. When it becomes clear that *no hope or dream can ever be realized.* As with the Palestinians and others. Thoughts of rationality and morality are gone. Now only thoughts of ending the nightmare and getting some revenge.

This is the result of politics at its extreme. Events such as *9-11* should be viewed as an inevitable result of politics. Especially of democracy. As this tragic event happened, I said to myself: "This is an inevitable result of democracy. Of a societal system based on coercion". Be a Homo sapiens. Think about it.

Galambos

Background

Most people do not recognize the name Galambos.[54] Those few who do fail to understand his achievement. They deprecatingly refer to him as a *libertarian*. He was, but for both of us that term has almost no meaning. It means too many things and none of them represent the true meaning of the concept. Of what it is to be a libertarian. I think it is a bit like saying a person is a "junk bond salesman". You will find the correct definition of "libertarian" in the definitions appendix.

The word *liberal* comes from the Latin *liber*. The adjective meaning "free" or "open". In the political sense it means "free" not a "slave". With the suffix "al", it means "pertaining to that which is *free* or *open*". With most Americans and others who profess to be "liberals" it means the ability to take things away from one person and give them to another person. If necessary, using the police powers of the political state to do this. This bears no relationship to the concept of being "free".

I will not write a general type of biographical sketch. I will only state a few basic facts about Galambos. It is his achievement that is important and that is the subject of this chapter.

Galambos the individual

Galambos was born in 1924 in Budapest, Hungary. His father—a successful architect in Budapest—knowing that World War I had not solved anything and that the conflict would soon continue—moved the family to New York City. He did not want his son to become cannon fodder in the next war. Galambos did serve in World War II, although he did not see combat.

His interest was in astrophysics and he received a Bache-

---

54 Andrew J. Galambos (1924-1997). There are a few other prominent scientist with the same last name, which in Hungarian means "pigeon".

lor and Master's Degree from Carlton College in Minnesota. It was there that he also met his wife and best friend (one and the same person).

He later taught at Claremont College in California and worked in the aerospace industry in California. In the 1950s space flight was still part of science fiction. Galambos was one of the few people in the aerospace industry who knew anything of rocket flight dynamics. Although his contribution was forgotten because people in the industry were thought to be part of a "team", much of what is now known was the work of Galambos.

From his father Galambos learned what he would later refer to as "positive history". True cause and effect in history. And from his studies in physics—which he considered to be the basic science and the source of all useful knowledge—he learned the laws of nature related to interpersonal relationships. Cause and effect in what is commonly and incorrectly referred to as "social science".

The Achievement

As an overview, I believe Galambos' achievement was:

- To identify and understand the achievements of our ideological antecedents.
- To build on this knowledge using the inputs of these ideological antecedents. By—as Newton would have said—"standing on the shoulders of giants".[55]
- To affirm that the universe is comprehensible and the method for comprehending it.
- To introduce the concept of *rightness* into the Scientific Method so it can be used to create and validate knowledge related to the interaction of thinking beings.

---

[55] "If I have seen further than others, it is because I have stood on the shoulders of giants". Stated in a letter from Newton to Robert Hooke on February 5th, 1676.

- By evaluating the moral content of action taken pursuant to a hypothesis you can then determine if knowledge is *right* on an absolute basis.[56]
- To recognize the need for absolute definitions and semantic precision in the creation of new and useful knowledge. And that failure to do this is at heart of the failure of "social science". And a major reason we have never solved even one of our societal problems.
- Creation of the concept of the Natural Estate.
- A proper concept of paying royalties for the creation of value in society. Which includes the use of the concept of "remoteness dilution".
- Creation of the definition of life. This enables the creation of postulates for biology, thus moving it onto a firmer foundation.
- The determination of the purpose of life. Understanding this is essential to the creation of the solution for pollution.

He created many other things, but maybe this is a representative list of some of the more important things.

For me, Galambos stands at one of the two "focal points" in the *ideological flow-stream of history*. A concept he himself identified and on which he placed Newton at a single focal point of history. This placement recognizes Galambos as one of the two most important people who ever lived. The other one of course being Newton. Why do I say this?

Because for any thinking specie to survive in the long term they must integrate both the volitional and non-volitional sciences. The volitional sciences being that area of knowledge related to the interaction of thinking beings. What we generally and incorrectly refer to as "social science". But as practiced now it is neither social nor science. The former meaning it does not add to the quality of life. The latter that a

---

56 *Right*: "That which is based on true premises and valid and logical thought processes, and the action dos not result in *intentional* interference in the property of another individual".

proper definition of *science* is "*organized, useful knowledge*". Which does not describe the results of the current body of knowledge on that subject.

The non-volitional sciences are generally referred to merely as "science". Which does properly describe it, as it is organized, useful, morally-acquired knowledge. However *technology* might be a better description, although that is more the result than the body of knowledge itself. We could also say "inanimate science". But why do we not simply use "non-volitional sciences"?

People create *ideological flow streams* today. Referring to them as a "time line". About the only thing missing is that Galambos had many time lines, flowing into and out of a point. The point at which someone—such as Newton—gathered all the individual lines or inputs and sat down and thought about them. Then engaged in "discovery". As described by Albert Szent-Gyürgyi von Nagryapolt (1893-1985) when he said:

> *"Discovery consists in seeing what everyone else has seen,*
> *and thinking what no one else has thought."*

This actually describes the second step of the Scientific Method. The formulation of a hypothesis. Then all you need do is corroborate your hypothesis and you have a theory. Corroborating it against history and existing, known laws of nature. If the corroboration fails you make another hypothesis and try again.

There is no such thing as "political theory". This is an oxymoron. Politics is a series of failed hypotheses. It does not qualify as a theory.

It was part of Galambos' genius that he recognized the need to test any planned, intended or actual action resulting from a hypothesis for morality. To see if that action, if taken, would be moral. If so, he said the result would not only be correct but *right*. Thus elucidating the concept of *rightness* in volitional knowledge.

If we apply this concept to "social science" we see why it has failed. Not only is the associated knowledge not organized, but virtually any action resulting from it is immoral. When a person understands this it becomes very hard to understand why anyone would believe that any problem can be solved through politics. And why we continue to accept and to tolerate a societal system based on the concept. This thought alone would be enough to cause a person to sink into dementia.

I think Galambos realized that his concepts, while brilliant, elegant, intuitive and succinct, were still a first step approximation to the final, comprehensive set of laws and body of knowledge for the volitional sciences.

As I have indicated here, Galambos' own work depended on that of:

- Newton.
- Paine.
- Darwin.
- Bastiat and Morgenstern.

Galambos himself identified many others. Those who provided the inputs to the aforementioned giants. Those from other branches of physics. And especially the very fundamental laws of physics: Thermodynamics. It was this knowledge that Paine did not have because it had not yet been developed. Again indicating the accretionary, evolutionary nature of knowledge. If Hero of Alexandria and Archimedes had the tranquility to complete their work we might have had a fundamental understanding of thermodynamics—including entropy—two thousand years ago.

I think it might be safe to say that Galambos created the *Carnot Engine* of volition. Certainly the principles are there.

Are not Galambos' "Laws of Volition" a *Carnot Engine of Volition*? To minimize entropy in societal interactions? Entropy being disorder?

I have taken the liberty—had the temerity?—to rename these as the *Universal Laws of Volitional Behavior*. I know for certain that Galambos understood that the Laws of Volition were universal in nature. I also refer to these as *Galambos' Laws*. Which is rational, moral and proper.

As he said, the monumental achievement of Newton was a first approximation to a "System of the World". That Einstein with his creation of *Relativity* had created and provided a second approximation. And that there would surely be a third and perhaps last one. Since Einstein had searched for this himself but did not find it before he died. Now, almost fifty years later string theory—a possible solution—appears to be unraveling.

Galambos—as Newton—stands at a focal point in history. Bringing together the thoughts of the great "humanists" in history. Those who sought to understand the nature of man the specie. Homo sapiens. The rules or laws covering the interaction of human beings. To affirm that it is possible to comprehend what nature says about the relations between individual thinking beings. The summation of which is the society of Homo sapiens. To bring order to an area of knowledge that has never had order. As evidenced by individuals being at war with each for more than 90% of the time for the past 3,000 years.[57] And never having solved one of our societal problems.

Newton said the universe is comprehensible. And this must include that part related to one of its occupants, Homo sapiens. As we are part of this universe. Galambos looked at the agony of a thinking specie trying to understand itself. And failing to do so because it had stopped thinking rationally.

Part of Galambos' genius was in understanding the true nature of a thinking being. What motivates a thinking being. The dominant characteristic that determines how a thinking being will respond to interaction with the inanimate and animate worlds.

Galambos said that all discrete bodies of knowledge must have two or three postulates. In the sense of "something taken as a given" and not "that which is accepted without proof". Galambos would not accept something that could not

---

57 Henry Grade Weaver, *The Mainspring of Human Progress* (The Foundation for Economic Education, New York, 1997), 263.

be proven using the laws of nature. And would not accept that anyone else should do so. This would constitute the acceptance of fallacies. And lead to intellectual dishonesty. A concept that is at the heart of the reason we have never solved any of our societal problems. Almost everything we believe in what is commonly referred to as "social science" is false.

Postulates are something you can use to create new and useful knowledge. And to prove or disprove existing knowledge.

Galambos referred to the postulates he created as *The Laws of Volition.* I have had the temerity to rename them the *Universal Laws of Volitional Behavior.* And the gratitude to refer to them as *Galambos' Laws.*

The First Law:

All thinking beings seek to acquire property.

Think. Were this not true we would never have come out of the cave or down from the tree. This expresses our curiosity about nature. What drives us to create new and useful knowledge. Which is the first type of property a thinking being creates. Without thought and an idea there is nothing. There can be no action.

First Corollary:

Individual thinking beings seek to acquire the most amount of property with the least amount of effort.

"Hah!" you say. "That confirms naked, unrestrained greed in thinking beings". The antidote will appear in a moment. This is the engine of creativity and innovation.

Please think about these two statements. Volitional life would not even get started were these concepts not true. The opposite would imply complete lethargy and the inability to interact with our surroundings. If the above was not true we would never go into the cave or climb the tree. Let alone go out of the cave or come down out of the tree and start walking.

For more thoughts on this I would recommend *The Terri-*

torial Imperative, by Robert Ardrey.[58]

Second Law:

Rationally thinking beings seek happiness.

Again, please think about this. Would we get any place as a specie if we sought unhappiness? We did everything we could to be unhappy? I must admit that some times this does seem to be true. Especially as regards our societal system. We bring untold misery on ourselves through this system. Famine, war, poverty, terrorism, crime, an appalling educational system, a travesty on the concept of health care amidst great advances in biology. Inflation, stagflation, depressions, recessions, etc. The result being a societal system that—in millennia of trying—has not allowed us to solve even *one* of these problems.

As individuals we seek happiness through the pursuit of our individual hopes and dreams for our life.

And happily this is the law that "regulates" the apparent capacity for unrestrained greed implicit in the First Law.[59] The moral regulator. As opposed to the coercive, immoral regulation of political laws.

Societal Concepts

General

The Natural Society is a simple concept. Since it needs only the laws of nature. Which already exist. As they have since the universe was created. Of course our elaboration and understanding of them has taken some time. As Homo sapiens sapiens we have been working on that for at least 140,000 years? Although we started laying the foundation about 2,500 years ago, it has only been in the last few hundred years that we have made much progress.

---

58 *The Territorial Imperative: A Personal Inquiry into the Animal Origins of Property and Nations,* by Robert Ardrey, Kodansha International, 1997

59 But it is a moral form of regulation. Depending on the individual being selfish and acting in their own best interest. Wishing to do only those things that bring long term happiness.

I have already described the really basic laws or concepts. Such as those elaborated by Scipio, Newton, Darwin, Paine and Galambos. And from these we can understand what are perhaps more appropriately labeled as "concepts" rather than laws of nature. Although being based on the laws.

Job Security

Background

Whether in a democratic society based on coercion or a free society, job security is essential. And simple. It is the way to avoid poverty or to get out of poverty. The only moral way to do so.

There is a major problem with job security. In explaining it. It is such a simple thing, and people are inclined to reject that which is simple. Preferring something complicated and almost incomprehensible. Which seems to indicate it must somehow be "right".

You must think very deeply about the principle of job security that I will state later. If you believe you have heard it before, think again. If you think "it has been tried" and does not work, think again. It *does* work. It is the *only* avenue to job security.

Time on the job is almost always meaningless. More than that, it is a dangerous concept. It lulls people into believing they are "good" and are "needed". Unless that "time on the job" had made you about the best person in your industry at performing the associated tasks the time is meaningless.

Another dangerous thing is the value of those very tasks in the market. Maybe it is part of a dying industry. Or tasks that are currently being or will be "outsourced". Because the money currently being paid for them is not consistent with their value in the marketplace..

Accounting system

Something that confuses job security and employment is our accounting system. I will not go into this in detail, but ask you to simply think of a few things. And then extend that.

Whether an individual or a company, there must be "money left over at the end of the month". Individuals think of it as just that: money left over at the end of the month after all expenses. At the very least it should be "zero", and not negative. Businesses think of that as "profit".

Businesses have "sales". This is their income. The same as wages are your income. What you as an individual are selling is your ability to do a job. One that brings in income for the company. So actually you as an individual have sales also. Your product is the service you perform for your company.

A business uses "cost of goods sold". What it costs them to produce a product for sale. Usually these are "direct" costs. Such as wages, materials costs and other such things. But probably not marketing costs, research, legal, banking and accounting fees and other such things. As an individual you also have "cost of goods sold". Which might be a portion of the monthly car payments for the car you drive to work. The petrol, part of the maintenance costs, your lunch, clothes and other such things. These would form your "cost of goods sold".

By class or type of expense there is a great similarity between business and individuals. Just different names for the same thing. It helps if both—company and worker—understand this.

Principle of job security

Maybe I should lead up to this more, but I will state it here.

Job Security

> *The ability of an individual to perform a task or tasks currently in demand in the marketplace at an acceptable market price.*

As long as you can continue to do this you should have no fear of becoming unemployed. The better you do it the more money you should make and the less chance of becoming unemployed.

And if you should become unemployed because of company problems not related to your abilities then you should easily find a new job.

Maintaining job security

This is something that requires continuous maintenance. You must keep your eye on the market. Make sure that your skills are still in demand in the market. That this is at least stable and not heading down. If heading down it means the company itself may have future problems for their products.

Understand whether the market value of your skills is stable, going up, or going down. Even if very good, if there is something additional you might learn that will make things even better. Think of the emerging market for a new skill that might be similar to your own. A compatibility with your education, abilities, interests.

Maybe there is even "something you always wanted to do". But didn't. Why not? Why not do it? If it is reasonably within you ability to do so. You must be flexible in your thinking and not feel you must always do the same thing. I have worked in radio, radar and communications, then in computer systems. Within that, aircraft computer systems, business applications and then process control computers. Then with disposable medical devices, both the manufacture and sales. Which included the creation of a foreign manufacturing facility. And then in the aviation and airline industry. Still later, the development of computer database systems accessible from the web.

And from time-to-time thinking about our strange, bizarre societal system. And searching for signs of terrestrial intelligence. I am relatively certain there is extraterrestrial intelligence. I am less certain if there is terrestrial intelligence.

The thing that really makes this concept of job security work is a proper concept of production compensation. Which will be discussed following.

Summary

You think I have said too little? Not explained it enough, with more examples? I am depending on you to *think* about the *principles* of what I have said. If you do, then you will understand it. And you will be successful. If you do not think, then you will not understand it. And you will not be successful. I am not "talking down" to you. I am talking up to you…

Production compensation

Background

Again, a very simple subject, concept and principle. Probably a major reason it is not considered. As in fact it is the key to profits, sales, cost control, productivity, innovation, control over individual life. And many other good and important things.

The principle

This is another one of those "why didn't I think of it" things. It should be obvious that it is the only rational—and moral—way to compensate an individual for satisfactory task completion.

Production Compensation

Production compensation must be based solely on the market value of error-free task completion.

Innovation is the engine of productivity. And this principle also applies to innovation compensation. Which means the innovator must receive a "royalty" or fee based on the sale of products that include their innovation. And for as long as that product is sold in the marketplace. Even when the innovator is no longer associated with the product or the organization that is producing and selling it.

If you create an innovation—a product or process improvement—and that earns the company, say, an additional $1 million in sales, you should get a percent of that. A notional figure I often use is 5%. Which means you should get $50,000 per year if sales stay the same. And not perhaps 10¢ more per hour. For company accountants who scream "this is too much", I ask they think about the ROI. Return on investment. Which is $950,000 in one year on a $50,000 investment. I think that is very good, how about you?

In fact, almost too good. Maybe you should get 10% and $100,000. The company would still have a very good ROI.

When this principle is implemented, many good and important things happen, as I stated above.

This form of production compensation means that in principle "worker" income is unlimited. It means automatic cost control for the company. It means both worker and company are each in control of their own destiny. And the "worker" is no longer a worker. They are in fact an entrepreneur. Although I more often refer to them as a "production associate". With an entrepreneur being someone who contracts with a number of production associates to do many different tasks and jobs.

If your employer is not using this concept now you might—after you think about it more—propose it to them It could ensure your continued employment. As long as you truly understand it and explain it properly.

This is the subject of a forthcoming book, *Productivity Galore*. Which I hope can be released in 2007. This will of course have much more detail on the concept.

Personal security

General

The intelligent ruler keeps his victims happy. So that—as Huxley said—they enjoy their servitude. It makes it much easier to rule them. This is done by trying to take care of most of their potential worries. Through unemployment compensation, healthcare, a pension system, etc. But since these programs are basically flawed, they do not work well. However, it is very easy for you to create a very good pension system for yourself. One that will leave you with significantly more money. And with it being *all yours*. Totally under your control.

I am talking about a plan that can take most of your life to bear fruit. And I have been saying that we might all be extinct in 30-40 years. But we will not be if we think about our problems rationally and work on them. And if you prepare your own pension plan—have nothing to with any of the systems of the régime—you will feel better about yourself. And when we have conquered the problems of Global Warming and Global Dumbing, then you can truly sit back and enjoy life.

The system

There will be some repetition here. Repeating something from the discussion on Darwin. Although quite short, it is very important. Because it can lead you to financial independence later in life. As understanding what true *job security* and *production compensation* can do for you now. These three concepts of job security, proper production compensation and your own very good pension system can do wonders for your life. Both short and long term.

And it can do this without interfering in your current state pension system. For as long as that continues to exist. Let them go on taking your money. Eventually it will stop, but you must not do anything that will further disrupt order in society.

Compare something very positive. That would end the problem of state-controlled pensions. Invest $10/week in an average performing mutual fund or other investment. Something earning, say, 12% per annum over a forty year period. Don't think about what is happening to it, just do it.

When you visit your account after forty years you will see that you have more than $400,000 in it. In terms of what today's current value will do for you. Which, if there is still some inflation in the future will be more in monetary amount. The $10/week is much less than the state steals from you for their bankrupt system It originates from the 1970s, when the *maximum* they took was about $500/year. If you can imagine that? And the $400,000 is much more than will be in your purported account with them. And you can use it as you wish. Leave it for another year and you may have another $48,000 in your account. That is 12% of the value of your account.

Okay, so with politics next year is a bad year. But leave it for four or five years and things have recovered. And now you have $200,000 to $250,000 more in your account. At least $600,000 to $700,000.

You question whether or not this is a viable possibility? Are there really investment vehicles that return 12% per year over long periods of time? Check the results of a large number of mutual funds. Or "unit trusts", as I think the Brits call them. There are many who have averaged 12% or more over a forty year period. Will this occur for the one you select? Maybe, but not guaranteed. You must investigate and make the best possible choice for yourself. You may also want to use one fund for, say, ten years. And then start another. Perhaps putting $10/week into each. As you make more money the additional $10/week will not really be a burden. And you diversify while also leveraging your investment.

"But your figures are not realistic. You totally disregard taxes." Yes I do. Everything is taxed and of course the amount varies from time to time and individual to individual. But also remember that when you start withdrawing money your tax situation will be totally different than it is now. And if we have been successful in eliminating Global Dumbing there will be no such thing as taxes.

If there is, we have lost. Yet to answer lingering questions: you will be old and retired, and your tax situation should be different and better than it is now. Also, over the years the additions to your fund may be deemed to be "capital gains," with lower taxes than ordinary income. I will not talk about so-called 401K schemes, or however they are called. I am not offering investment advice other than that you should consistently invest in your future over a long period of time. Just that simple principle.

This investment fund should never be touched during the forty year or so period. You must be very strict with yourself about that. If you start making exceptions they will expand. I suppose we must say that if it becomes a matter of life or death, then you may use money from the account.

But, depending on how much you can push yourself in the beginning, you might establish an "insurance" or "medical fund". And you have such an expense that can not be handled any other way, then use this fund. If you did not have any extraordinary, uncovered medical expenses during the first twenty years or so, you could become self-insured. And you put the money you were paying for medical insurance into your medical fund.

Remember also, it might make sense to borrow money to cover your current problem rather than using your fund. Why? Because you may be able to borrow at a much lower rate than you earn on your mutual fund account(s). Plus, this forces a certain additional discipline. You will be very careful how much money you borrow and how you use it, since there is an obvious cost. The cost of tapping your mutual fund could be quite high in the long term, but much less noticed.

It is becoming increasingly hard to write more that is of value. It is really such a simple subject and concept. So easy to do. Put the $10 in an envelope every week. Then about once a month put it in your bank account and write a check to the mutual fund. Or transfer it electronically.

One thing you should think of, so you do not become depressed in the early years: the amount in the fund will grow very slowly. Look at the table:

Year	**Amount**
10	$9,125.-
15	$19,000.-
20	$37,000.-
25	$69,000.-
30	$125,000.-
35	$224,000.-
40	$399,000.-
44	$630,000.-

For almost ten years it seems as if very little happens. But you should not fret about this. And hardly even pay any attention. Maybe every five years look to see if the return is averaging about 12%/year. Taking into account whether the years were "average, good, bad", etc. You want to look often enough and close enough only to see if your fund is meeting its general objectives. And if not to perhaps start putting your $10/week into another fund. Or leverage things and continue with the old one but put another $10/week into a new one.

Notice that in the beginning the $520/year you put in is what creates the growth. But slowly this changes. In the latter years you could stop adding that money and not have much effect on the end result. At thirty years your interest earns about $15,000 or 29 times the $520.

The reason for the 44 year period is assuming you start working at age twenty-one and work for 44 years, you are then at the magical age of 65. Still a general age for retirement.

And please, if your numbers turn out to be different than mine, don't quibble about it. In a sense, if you end up with $400,000 or $700,000 it doesn't make that much difference. It is vastly more than you get by doing nothing. And as I say, if you have the $400,000 and a good investment vehicle, the next year you may have about $450,000. When the value of your account is large, each year can bring much more.

Think of another thing: If you wait until you are fifty to do it, you will end up with less than $20,000. Or if you want to be where you should have been and could have been, you will need to dump in up to $125,000 the first year...

Social Insecurity might bring you $1,000/month. Maybe even $1,500. But with, say, $400,000 in the bank you could take your dividend/capital gains check and have about $4,000/month. Without denting you earning power. And if you start early and continue until sixty-five you can pay yourself more than $6,000/month. Which might be more than you ever earned in your life...?

Individual Freedom & Responsibility

I know I am repeating myself, but I think it is necessary. Perhaps the phrase and concept of "individual freedom" is overused. People are blasé about it. "Yeah, everybody says that". Or "You are in favor of it, so what? We have it. It is part of the American system". Things like that.

But Americans do not have it. It is not possible for it to exist in a society based on coercion. Whether the style is that of a democracy, dictatorship or monarchy. Or economic variants of communism or socialism. All democracies are socialist, increasingly trending toward communism.

Yet individual freedom is the natural condition of thinking beings. It is a prerequisite for successful volitional (thinking) life. Without it competence—fitness—declines and eventually disappears. And with it our specie.

It is not a danger as most believe. People seem to believe that if there is individual freedom with everyone doing what they wish that there will be anarchy. Anarchy is something that can exist only in political societies. In societies based on coercion. It and its companion, disorder. Which is an evitable result in societies based on coercion. But absent in mature societies based on individual freedom. I do understand that in a transitional society, between coercion and true and complete freedom, there would be some disorder. Although not nearly so much as now. Do you think things could really get worse than they are now? Did you ever believe things would become as bad as they are now?

True and complete freedom without disorder depends on the companion of freedom, individual responsibility for the result of action. Aided by an understanding of the credit mechanism. With the need for that simple concept and its efficacy being one of the great insights of Galambos.

That and the fact that selfishness—individual self-interest—is a desirable trait. A necessary one. Part of the Universal Laws of Volitional Behavior. Part of the almost involuntary drive for happiness. Because it is part of your self-interest that you do not intentionally take action that degrades the quality of life of another person. Because by doing so you will create disorder in society. In the form of depression and possible retaliation for your action.

Individual freedom is a prerequisite for successful volitional life. You say, "But yes, ok, I agree with you. We have never had true individual freedom on this planet. Yet we have survived". We have survived until now. But the game is not over yet. And while we have survived until now, we have not prospered as a thinking specie. Unless you believe that the daily news—including the increasingly breaking and broken news—means prosperity. That these events are the best that life can be and will be.

Think of the hopes and dreams for your life. What you want from life. Know that family members, friends and all other thinking beings also have hopes and dreams for their life. And that there is *no possibility* these can be realized without individual freedom. And the reason we have chaos and disorder on this planet is because we do not have individual freedom. That this is becoming a terminal condition.

But that by understanding our true nature—the true nature of thinking beings—we have hope. We can reverse the decline. The way is simple. And the need is there. I have said that "regulation" as is performed under politics is immoral. And irrational. As it has not and can not work. I have also said—as I learned from Galambos—that there is a moral form of regulation that does work. We see a hint of that in our current, defective societal system.

Regulations

I suppose I have this heading as an introduction to the following one on the credit mechanism. And that I don't really need this. But maybe I want to use the term and concept one more time. To bid it farewell. Or to resurrect it. Almost a miraculous resurrection.

Considered against the backdrop of politics it is an odd form of regulation. It almost seems like a market product. Which is what it is. In the sense that accounting is a market product and service. To know where you stand financially as a company or as an individual. In a rational and durable society your credit rating tells you where you stand in society. Almost saying whether you exist or not. Because if your credit rating is not acceptable, you really would not exist in society. You would be irrelevant in society. A non-person.

If you appeared at the supermarket with your trolley for checkout the clerk—robotic or otherwise—would not even see you. It would be "next customer, please".

This sort of regulation replaces the one that says "Obey or else". And the political police, political court system, political prison system with executioner. All costing a huge amount of money. Not preventing crime, not bringing justice to anyone.

The credit system as a regulatory mechanism is so much a part of nature that—as nature in general often seems—it is invisible. It just works. Yet if you ignore it by trying to contravene it, nothing works in society.

The beauty of nature

I decided to take a break, and returned to the bedroom. Where my little daughter was still asleep on the bed. I looked at her and thought of two cells coming together and producing billions. Creating two kidneys, a liver and gallbladder, spleen, lungs, heart. A vascular system. How does it do that?! And then, "suddenly", charging it with blood. Making that wonderful brain. How does "nature" know to do that? Maybe it has "intelligence"? Not in the sense of what religious mystics believe.

Using computers we strive to create "artificial intelligence". But maybe nature has a form of "natural intelligence"? It has "wired itself" and *randomly* (not consciously!) "programmed itself" over billions of years. As a neural computer does. And as nature does—and we should do—if something does not work it is "discarded". Falls by the wayside.

The big difference between that and our computers and

attempts at "artificial intelligence"? The "clock rate" of the computer. Its time to perform what is in effect a very complex calculation. Sometimes taking millions of years. And it would seemingly have a much larger storage capacity? Perhaps approaching infinite?

And nature is its own programmer? It has shown the ability to learn and to adapt. It does this by passing the results through a "filter". The inanimate laws of nature. As indicted by the results of its "computational" processes. We and all the other life forms—and non-life forms—are the result of the actions of this processor we call "nature".

And like the computers constructed by Homo sapiens, this Natural Processor sometimes makes mistakes. Although as with our computers, it must be a problem with the programmer, and not with the processor itself? Yet, as I say, it learns from its mistakes. And is able to use its own form of "artificial" intelligence to "re-wire itself" and produce something that does work.

The Credit Mechanism

With credit to Galambos, this is the moral regulator in a rational and moral society. This concept—that the credit rating of an individual can act to regulate behavior—is one of his many great achievements. It is almost like that cherished institution (as long as their rating is high) of the politicians: the poll-taker who provides an opinion on your life. And is very similar to our current credit systems. The main difference with this is that the "credit rating" is more for the societal behavior of the individual than their personal financial worth. Yet quite like your credit card rating.

In a society in which each individual has 100% control over their own property, all discovered property would have an owner. Unlike now, where the political state claims they "own" property. But actually only control it. As ownership is a moral concept and can not be acquired through force or fraud.

If this was really true—that taking control of property by

force or fraud conferred ownership—it would make a shamble of society. Which of course is our problem now. As with this concept it would mean I could steal your car and thereby become the owner, yes? Rather than simply a thief who is controlling the property of someone else without their permission. As it is, politicians reserve this special right of theft for themselves. Another of their monopolies.

This exposes the fraud of the concept of "public property" for what it is: fraud. Because if I am an owner of Yellowstone National Park, I wish to sell my part of the property. As there is a Supervolcano lurking under there, and if it "blows" the value of my property will slump. Probably even "tank"?

A critical, important characteristic of ownership is that the owner may sell their property. Or refuse to sell it. And they can establish any price they wish. Too low and it may sell quickly at below market value. Too high and it will not sell.

As I said, in a moral and rational society everything would have an owner. Be "privately" owned. Although this modifier will be unnecessary when there is only "property". When there is no such thing as "public", régime-controlled property. This means electricity, gas and water systems. Which are now completely controlled by the politicians. Road and street systems. Also controlled. Schools. Even the "private" schools are controlled, but they do usually offer an education else they will soon go out of business. And, the "public" schools, well, you know about them…

What might be called "consumer stores" are nominally and effectively under the control of "private" owners. Including food stores. Why do I say this? If you were thinking about the subject the "light" might have come on.

As when I was a child, many shops had a sign that said "We reserve the right to refuse service to anyone". Unfortunately, much of the time this was done for an irrational reason. To refuse service to people for reasons such as skin color. Perhaps even religion if they could make a guess at this.

It is moral to refuse service to anyone. If it is your property you can refuse access to it. But perhaps not rational as you are turning away a customer and potential sales. Which would reduce your profits. Yet, as I said, it is moral. While to force you to accept anyone as a customer is immoral. And also irrational. Think about this.

What if, as in the section above, I steal your car? Or perform some other immoral act. And I come to your supermarket to buy food. It would be moral for you to refuse service to me. Also rational. Further, it would be moral and rational for you to report me to your credit rating bureau. So that other merchants would know I was not a trustworthy individual. And my electric, water, gas and telephone companies might cut my service. Lowering my quality of life. As happens when your credit company suddenly cancels your card.

How could I get myself out of this situation? And how might you receive justice? Such as getting your car back. And being compensated for all the problems I have caused you? Simple. And it does not require political police, a political court system, a vast system of political prisons. Euphemistically called "correctional institutions". Anyone know of anything this system has corrected? Other than the problem of unemployment for the prison staff?

After I was tired of being without electricity, gas, water and a telephone system… And the vegetables and fruits in my garden were nearly finished… what I would do is seek an appointment with you.

So that I could abjectly apologize for my moment of madness in taking your car. And return it to you. After first having it serviced by your local authorized dealer. Cleaned and "detailed" and so on. And said "Please, I would like to offer you money to compensate you for the problems that resulted from my irrational and immoral action".

Then if I was really clever I would ask you to please give me some guidance in this matter. An estimate of what it had cost you. Knowing you to be a basically kind and reasonable man. And what it would cost me might turn out to be less than I expected. And if I had made an offer of restitution it might cost me more than I need pay.

Assuming we reached an agreement—which we must really, else I could not get on with my life—I would make arrangements to pay you. You would have your car back plus restitution for your "out-of-pocket" costs for my caper. It would then be your obligation and duty to report the event to your credit rating service. Which would then remove me from the "black list".

But some will wonder what might happen if you were not the nice person I thought. And demanded a huge amount for restitution. Something I could not or would not pay. What then?

The credit system would have an "arbitration" service. It could be a separate company. And I could explain the situation to them. Explain what I offered and what you demanded. I could pay the restitution into an "escrow account". And the service would visit you and discuss the matter.

If you then accepted, fine. If not, and they felt the payment was "fair" and "according to the market," the credit bureau might restore my credit. With the case history available to anyone who wanted to see it. And you might get a notation in your record saying you were unreasonable. This should not cause you a big problem. Unless there were many such cases. And it might help to lend credibility to anyone who accidentally interfered in your life and you sought to have them "black listed," claiming it was not accidental.

So the system can work to accommodate different situations. And work much, much better than the current system.

You say "this is too simple and unrealistic. It will not work". Does our current system work?

I recently read of life in a "small" women's prison in central California. With more than 11,000 inmates. I am probably correct in saying that none of the victims of these prisoners has received justice, yes? And "society" has not received justice either. Many of these prisoners will come out and commit a crime again. At the very least the crime of being a burden on society. Exacting a huge amount of money from society to operate the non-functional "justice" system. Plus everyone in that political entity must pay to support the failed "correctional" system from which no one receives justice.

Please think…?

Chapter 6

Natural and Societal Disasters

Synopsis

The basics are things such as rising sea levels, temperature problems and rainfall patterns,. Rising sea levels are almost a simple problem. The sea levels will rise, and that is it.

High temperatures are of course the driver for rising sea levels. But high temperatures can cause other problems. Desertification. The rain forests spontaneously bursting into flames. There can also be low temperatures and ice storms. Changing rainfall patterns can also produce desertification. And flooding. Many very unpleasant effects.

But as *Global Dimming* was the problem hiding the true dangers of Global Warming, so is *Global Dumbing* hiding the problems related to our societal system. And this is far more dangerous than Global Warming itself. Until Global Dumbing is fixed we can not hope to solve the problems of Global Warming…

Preface

The problems are two-fold. The first problem is the actual environment. What we might call "technical" problems. What happens to our environment in the sense of rising sea levels and temperature changes. And within the latter things varying from cold to hot. Freezing weather and ice storms to drought and desertification. Floods to drought.

The second is the effect on society. The disruption of trade and the food chain. The effect on employment, the quality of life, the dislocation of populations and associated things.

I intend trying to give only an overview of the former: the general changes in our environment. Climatologists and related types are far better qualified to provide the detailed environmental information. Further, that is very much of a dynamic thing as we learn more about how to construct the "model".

For the other, the principle of what will happen in society is unchanging as *principles* should be. And we must talk first in principles else we will not solve our problems.

I will try to discuss the societal problems only in principle. I could provide details and have written something on that. But I have learned through thirty or more years of bitter experience that people are not very much interested in reality and the truth of their societal system and situation.

As Aldous Huxley said in *Brave New World*, "… since reality, however utopian is something from which people feel the need for taking pretty frequent holidays".[60]

Most individuals are influenced by group thought and are conservative. Not wishing to entertain new and potentially useful ideas, but in conserving their societal fallacies. Including those related to their culture and religion. And ignore the consequences of doing so. As Thomas Paine said:

"A long habit of not thinking a thing wrong
gives it a superficial appearance of being right…"[61]

This is the reason that virtually all of our so-called societal knowledge is wrong. Riddled with fallacies and unusable. It has never been tested for rightness. To see if its application is moral, rational and will increase the qualify of life. Instead, it has unthinkingly been accepted for millennia. Even as it degrades the quality of life for all and causes chaos and disorder in society.

It is necessary to consider from where it is you receive value in life. Whether it is from what is usually referred to as your "government" or from elsewhere.

To analyze your thought processes when you have de-

---

60 Aldous Huxley, *Brave New World* (Perennial Classics, New York, 1956), Foreword, xvi.

61 Thomas Paine, introduction to *Common Sense.*

cided to purchase something or to subscribe to a new service. Compare the result of what you get to that promised for the alleged government "services". A service is a product: it is simply an intangible product or less tangible product. Or it may be tangible in the sense that it provides a telephone device for you to use when you wish to talk to someone.

Something to make the light bulb produce light when you flick the switch. A road to drive on when you wish to go somewhere. A school for your children.

When purchasing a product, among other things you may think of various sources of supply. Of the cost factor and service. Before purchasing you may think of whether it will truly add value to your life. Commensurate with what you will pay.

As my teacher said, when you purchase something there is really a disagreement rather than an agreement. If it is a car, you are valuing the car more than the $10,000 you will spend for it. The seller puts a higher value on having the $10,000 rather than the car.

So you should feel that what you will get in exchange for the money that leaves your account and is no longer with you has less value than what you acquire for the money.

When you purchase an Apple *iPod* you feel that you would rather have the *iPod* than your $200 or so. When you effectively pay $500 per month in taxes for the school with the metal detectors, 30% or more of whose "graduates" are functional illiterates, you need to think if you are receiving value in life.

Or when $200/month goes to purchasing part of a laser-guided missile for the Israeli's to use in Lebanon. With the cluster-bomb option or without it. But Bush wants to supply the ones with a greater, wider explosive power and the cluster bombs. So you may get less for your money. Although in the end you are destroying more property and killing more people. Which is what the ruler wants, yes? If not, why do it?

If you pay $50/month for electricity you are getting something of value for your money. Although you must use the state-controlled monopoly selected by, well, the government, yes? The people who are against monopolies.

Some thoughts related to from where it is you receive value in life.

Background

I think all mainstream climatologists now agree that "Global Warming" is happening. And that it will result in rising sea levels. What is not known exactly is when the seas will rise and by how much. That there will also be climate changes that will drastically change the landscape and environment.

Changing areas of lush vegetation into deserts. Grassy plains being plagued by ice storms. Changes in or ending of monsoon seasons, plunging areas plagued by rain into deserts. Major rivers of the world disappearing. Rain forests spontaneously bursting into flames. Which of course will release yet more CO^2 into the atmosphere.

I think estimates of the amount of sea level increase vary from about three meters to sixty meters. A very wide range, and even the lowest estimate will have potentially catastrophic results for the civilization of Homo sapiens. Including the survival of the specie itself. The time period varies from perhaps as little as ten years for a several meter rise to maybe a hundred years for a sixty meter increase. With the ten year figure requiring a "tipping" event.

Even the lowest figure would probably flood the *Führerbunker* in Washington. And "the one without whom we could not live" would need to make a decision as to whether to stay there or abandon it for the "Wolf's Lair" in the Colorado Mountains.

The timing depends on whether there is a "threshold event" or not. Something that triggers a very rapid rise. Such things have happened in the past. They may or may not happen soon.

But I think we should assume such an event *will* happen and make preparations for the results. I think we *must* have the plan essentially completed and in use within five years.

The problem is that if we do not act quickly and properly—meaning dispensing with coercion as the basis for our societal structure—we may easily get ourselves into the situation in which we have run out of time. *And nothing we can do will avoid us becoming extinct.* And dying painfully, not in luxury.

Rising sea levels are not the only consequence of Global Warming. There is also the desertification of currently fertile areas. The increasing frequency and severity of hurricanes, typhoons and tornados. And currently temperate areas experiencing debilitating ice storms. The Amazon rain forest and others bursting into flames. As with the increasing forest fires in North America and other places.

Other than that, there is not much of interest to report on Global Warming. Well, maybe one small thing. If the ocean temperatures rise too much the methane hydrates will start to release ten thousand billion tons of methane gas.[62] Then it will be all over for Homo sapiens and all except some hardy bacteria and viruses. As the temperature on Aristarchus will spiral out of control, reaching levels not seen for four billion years.

It is better not to think about pension and healthcare problems. Killing each other in war. Whether to use bombs that cause a lesser or wider area of destruction. But to think about disabusing ourselves of a societal system that allows these things.

Tipping points

The famous "tipping points" are things that could cause a very rapid, sudden change. Things that have happened in the past and may happen again.

About 8,200 years ago, a very sudden cooling shut down the Atlantic conveyor. As a result, the land temperature in Greenland dropped more than 9 degrees Fahrenheit within a

---

62 <http://www.bbc.co.uk/sn/tvrado programmers/horizon/dimming_trans.shtml>.

decade or two.[63]

Rising Sea Levels

Ports

Environment problems

It should be obvious that ports are located where the land elevation is low. At nearly sea level, yes? And preferably a protected area. Perhaps a natural barrier against the tides and storm surges. And these will be the first victims of rising seas.

You may know or can look at an Atlas to see the ones in danger. In the US it is Boston, Long Beach (Port of Los Angeles). San Diego, New Orleans and others. In other parts of the world it is Hamburg, Bremerhaven, Rotterdam, London, Southampton.. Many places. The port of Cape Town in South Africa may be a wonderful exception.

Ports are at or near sea level. In the days of smaller sailing ships and at the beginning of steam with smaller ships the dock itself was not much above sea level. Some docks were floating, so the relationship between the main deck level (often the only deck) and the dock was always the same.

As the freeboard—the distance from the water line to main deck level—was not much. Today with huge container ships the freeboard and height of the dock above sea level is much greater.

But this is expected to be a constant or a standard amount?[64] Ships are usually designed with a certain freeboard to match a relatively standard dock height. And this is not expected to change much. But sea levels will rise and put docks under water.

Ports can become non-functional. Not viable as a port.

---

63 Peter B. deMenocal, Lamont-Doherty Earth Observatory, Columbia University. <http://www.washingtonpost.com/wp-dyn/content/article/2006/01/28/AR2006012801021_pf.html>

64 An assumption on my part which may not be completely true. Perhaps the height of the crane above sea level can handle greatly varying freeboard situations?

Unless there is some advance thinking and innovation. To provide a flexile solution for continued use of a port at any expected sea level rise.

With good advanced planning a port can continue to function. It is better to plan for that than to have a port slowly—or quickly—go into decline.

Societal problems

If the port continues to function then there will be relatively fewer societal problems. Except: what about the port city itself? The problems that can occur are:

- Population dislocation. With associated property value problems.
- Unemployment.
- Personal and company financial problems. Failures for individuals and companies.
- Electricity, drinking water, trash collection.
- Disruption of road, rail and bus services.

Coastal areas

Environmental

Over 80% of sandy beaches are eroding.[65] And 10% of the world's population lives in coastal areas that are less than ten meters above sea level.[66] That is 640 million people. Wetlands are becoming flooded, with the sea attempting to make new wetlands in previously dry areas.

Many coast cities in the US will be rendered uninhabitable. Certainly the most charming and historical areas of these cities. Their sea-level location made them a natural destination and town for colonial America. Look at elevations along the coast from New England to Texas.

I think there is more than 5,000 kms of coastline. For at the very least an average of 15 kms inland the land is at risk. That is about 80,000 square kilometers of highly populated

---

65 http://www.floridaenvironment.com/programs/fe01002.htm

66 http://www.sciencedaily.com/releases/2006/05/060517175614.htm

and very expensive land gone. How many people? Certainly a few million.

But there is a "hidden" danger. At least I think it is not well known or considered. There are some major towns inland from the ocean that are at or very near sea level. Levittown, for example is essentially at sea level, but with intervening higher land levels between there and the ocean. But what happens if a small rivulet of salt water finds its way to Levittown? It could soon become an uncontrollable torrent. And Levittown a salt water lake dotted with islands between there and the ocean.

Trying to fix such a place might be like the boy putting his finger in the hole in the dike. While you have your finger in the hole another one develops elsewhere. If you build a levee or dam to keep the water out then you will have an "ocean" dotted with islands between there and the main ocean.

Places such as Boston with its elevation of about six meters. With a very small sea rise, the "Big Dig" should fill with water, yes? I am not sure that building a levee would work any better in the long term than it did in New Orleans? Which itself will almost certainly become a place of history. Like Atlantis.

Philadelphia is in worse danger with an elevation about half that of Boston. And of course Miami at about the same three meters. Much of the old part of my long-time favorite city of San Diego will be lost. Houston has a lower elevation than Philadelphia, at about twelve meters.

Washington, DC is less than eight meters. And parts of the *Führerbunker* in the Whitehouse are below sea level. Alexandria, Virginia—the home of the CIA, NSA and other hallowed institutions—is only a meter above sea level. But with their "look ahead" philosophy and understanding of the world I am sure they will make plans to continue "serving the people". And with these comments they will almost certainly move me to the top of the eligibility list for a vacancy at Guantanamo. And Bush will pen an exception to not sending a native-born Americans there.

Hamburg is about half the elevation of Houston. And no stranger to disastrous floods. Although that was—I think—more from storm surge. The future one will be first from storm surge, but then also permanent.

It's companion cities of Bremen and Bremerhaven—while at twice the elevation (about twelve meters)—is not really much better off.

New York City at almost twenty meters may last until into the next century. When both the Greenland ice cap and the Polar ice has melted. Then of course it will be gone. Perhaps before, as the vast subway and sewer system presents a risk. The problems of the sewer system, electricity and all the stuff underground.

Major, historical parts of San Francisco at around twenty meters are also vulnerable. The future of The Netherlands is very much in doubt. I might say it is bad that they are not still associated with Spain, with its generally higher elevation. But many of Spain's coastal cities will have a problem. More than that, it will become very, very hot there. Temperature being in a sense a different problem.

And London. A city with which I am very familiar will—sadly—become a big problem. The "Docklands" is still almost in its infancy as a living, vibrant part of the city. But if the Thames backs up for many miles, much of London will have a very bad problem. And again there is the subway. It is vital to the viability of London. I don't think water taxis are a viable alternative.

Unlikely they could make it compete with Venice. Which will finally lose its long-time battle with the sea. Of all the costal cities of the world, Venice holds a special place for me. Going there often in the late 50s and early 60s. And feeling a great empathy for its past. It was one of the early places in Europe to receive the Latin texts newly translated from the Arabic. Containing the "lost" knowledge of the Greeks. Venetian surgeons were in the forefront of medical knowledge and practice in Europe beginning in the 1200s.

Much of the long-suffering Indonesia and Philippines will have major problems. The increased sea levels will add more load to the tectonic plate joint along the "ring of fire". This could cause more or stronger earthquakes and the resulting tsunami events.

A small rise in sea level can produce a dramatic effect on a coastline that is relatively flat. Where the angle of descent is very small. Such as the Eastern Seaboard of the US.

Over the last 3,000 years the effective sea level rise has been about 1 to 2 mm/year. Based on tide gauge data. On the relatively flat coastline of the Eastern US a 1 *mm* rise will cause the shoreline to retreat by about 1.5 *meters*. This is almost five of the king's feet.

This means that if there is no change in the rate of sea level rise, the shoreline on the East Coast of the USA could retreat by fifty feet in only ten years. And since the Greenland ice is melting at an increasing rate it could be much greater.

The "handwriting is on the wall". But you better read it now, as it will soon be under water…

Societal

You built or bought a house on or near the beach as this is what you wanted as a life style. But the beach will disappear. You will see a low sea wall. And part of the view you came for will be gone. The seal wall will rise yet more and most of the view is gone.

In time, without a seal wall, nature would simply "move the beach" farther inland. And if there were wet lands might also move farther inland. And the balance of nature would continue. But we humans operate on a different time scale and have not observed and abided by that of nature. So, space permitting along the coast we will need to make changes. No space there and we may need to redefine what is important to us.

If we really use our minds and take action as individuals, not as the collective slaves and drones of the régime, we can change the march of Global Warming and its consequences.

And for the next generation—perhaps even ourselves—begin to recreate the world we cherished in the past. It is only a matter of creating new knowledge.

Maybe we need to think about a "pause" in *consumer* technological development? Saying that "for the duration of the emergency" a 30 Gb drive is all we need for our Apple iPod. And instead we channel our resources toward reducing emissions and taking greenhouse gasses out of the atmosphere?

We could do this by consumers merely saying this is okay. And that it is okay for a company in which they are a stockholder to divert their attention and resources to Global Warming problems. And such thinking will be an indication that we are solving our Global Dumbing problems.

On the East Coast of the US, in Indonesia, the Philippines and many other places all it will take is one small "leak". A rivulet of water from the ocean to an inland point below sea level and it will soon become a torrent. And unexpected inland areas will be flooded. With salt water, not fresh water.

Although this is a bit of an exaggerated exception, think of what is now the Mediterranean Sea. It was once a vast canyon area. A "dry lake"? And there was a sea wall at Gibraltar. But the seas rose and some water breached the sea wall. And in a massive deluge lasting a century, that canyon filled and became the Mediterranean.

Are there similar places on Aristarchus where this could happen now? With a 60 meter rise in sea level? Probably not on the scale of the Mediterranean, but perhaps large enough to become an "inconvenient truth"? With credit to Al Gore and his film of the same name.

Island disappearance

Some islands—including the countries on them—will disappear completely. Others will be flooded so far inland as to make them "inhospitable" to life.

For me this is personally very sad. As I lived in Bahrain in the Arabian Gulf for almost five years. It was one of the most enjoyable periods of my life. To think that its people will lose their ancestral home is almost unbearable.

There have been people living there for at least 5,000 years. It was once thought to be the location of the Garden of Eden. And was in fact the location of what was thought to be only a mythical place, Gilgamesh. Its actual existence on Bahrain was confirmed by Geoffrey Bibby in his book *Looking for Dilmun.*[67] I had the good fortune and honor to meet Bibby in Bahrain about forty years after reading his book. He was still working on a "dig" there.

You can go to a local library and look at a good Atlas in the "reference" section. One that has elevation contour lines on its maps. And research this subject. For me, the Internet has not produced the information I thought it should.

For the USA, the *Back Roads Explorer* "TOPO" maps is a fantastic buy. You get 18 CD-ROMs. You need only move the cursor over the screen and read the elevation where the cursor is. It only costs $50. I think you can read more about it at:
<www.nationalgeographic.com/maps>

I have created a beginning database with some elevation information. I will try to add to it, but can't make any promises. Also, it is hosted on a server for which I have no control over its continued existence. You may check and see if it is working, at:
<universal-salvage.org.uk/books/reader.php#sea>
Click on "elevation information". You should be redirected to that server.

Low-lying Countries

A one meter rise in sea level will inundate half of the rice land of Bangladesh. This will effect millions of people. Millions of refuges would be created as a result. But this will also be true of China, Egypt, India, Indonesia, the Philippines, and Thailand and Viet Nam.

---

67 Geoffrey Bibby, *Looking for Dilmun*, (Stacey International Publishers, 1996)

Belgium and The Netherlands have historically been referred to as *Les Pays-Bays*, the "Low Countries". Geographically speaking, having nothing to do with the characteristics of the people living in them. It will be a massive, very expensive, and up-hill battle for The Netherlands to survive through the 21st Century. There are more than 16 million people there. And another ten million in Belgium. Of course Belgium does have some high country. But this tends to be more sparsely populated than the major cities. And I doubt the people currently living a tranquil life there would welcome a few million intruders.

How do you move a country? Especially one as large as Holland? And also Belgium. A better question is "why do you need a country?" In fact you do not. It is a primitive concept that long ago reached its expiration date.

I remember when Indonesia was fighting to retain E. Timor as a colony. And the people of E. Timor were talking about "independence" or "federation". As if those were the only two choices. I thought, "Why not the unspoken, obvious choice? That of individual freedom". But now it is Homo sapiens that must consider the option. And the choices are slightly different. Individual freedom or extinction.

Look at a topographical map of Indonesia, Malaysia, New Guinea. Remember the pictures of the coastal areas of Indonesia seen in the news after the 2004 tsunami? All the coastal cities are, well, on the coast. Beaches with little slope. While the incoming waves of the tsunami were huge at perhaps ten meters or so, that may be the sea level rise in the future. And then the more frequent hurricanes or typhoon surges will be on top of the old tsunami level. Not only will old coastal towns and cities be gone, but what was once "safe" upper-ground inland will no longer be safe.

With our current, defective societal system this could of course help to provide "full employment". Think of the bureaucracy needed! Tens of thousands of political employees needed to issue orders and permits to move. The tearing down of old houses and buildings and new construction. All of which will require permits.

And of course a person will not want to move until all documents are in order. Signed, countersigned, authenticated, red-stamped. A guarantee the new land is available, properly documented and ready for the move. Perhaps the destination land confiscated from the current owner. Who might then decide to become a terrorist.

Money to be printed to facilitate all this. Using the time-dishonored US system of government approved counterfeiting. Referred to as "monetizing the debt". And such a debt it will be!

Instead of individuals simply thinking what is best for themselves and doing it. I know: it won't be easy. But necessity could be the "mother of imagination".

Concluding thought

Once the sea level rises that it is. It may go still higher, but it will not go down. It will take *tens of thousands of years for this to happen.* Once Miami, London, Hamburg, Shanghai, Amsterdam and other places disappear that is it. The only thing we can do is to try to prevent this. There is already a sea level rise that is certain. We don't know exactly what this is. And we can do nothing now to stop this.

But what we can do is prevent it from rising even more in the coming decades and century.

If for example this is five meters—already a devastating amount, but probably certain—we might be able to prevent it from going more than ten meters. But the five meters—or whatever it is—is not alterable.

Melting of glaciers

General

Although this is the prime-mover behind rising sea levels I have made a separate section for this. Not really to discuss it as the cause of rising seal levels, but rather the fact that it can be part of a "tipping event" or "threshold event". And thus add to the urgency for societal change.

About three-fourths (75%) of the planet's fresh water is stored in glaciers or ice sheets. Which if melted would raise the sea level by about 67 meters, or 220 feet.[68] It has been estimated that over 100 million lives would be effected by less than a *one meter* (three feet) rise in sea levels.[69]

The tipping event

There can be "lubrication" at the bottom of a glacier that causes its rate of descent into the ocean to accelerate. The higher velocity of the ice is thought to be related to higher temperatures causing increased melt-water which can penetrate to the base of the glacier and hence reduce the ground friction. Surges of ice streams from the ice sheet can also occur due to increased accumulation at the head of the glacier.

Further, this causes earthquakes. And if one was large enough there could be—in geological time—an almost "instant" rising of sea levels.

In a sense this acts as a "tipping event" or a "threshold event" that greatly increases the amount of ice reaching the ocean and melting. And sea level rising.

A current estimate may say that the sea level might increase by three meters in fifty years, but in fact this will happen much sooner. And then ten, twenty, thirty meters and more can happen much sooner than we anticipate.

The effect of a "tipping event" could also occur through ignorance. Lack of knowledge, of measuring ability and so on. Unlike with politics, where the problem has been here for 5,000 years with no solution, this ignorance of a few years is understandable. We are operating in new areas of knowledge.

In fact, this has already happened. As on 11 August 2006 there was the report that the Greenland ice sheet is melting three times faster than previously thought.[70] And now (14

---

68 http://www.sciencedaily.com/releases/2005/07/050712233245.htm

69 http://www.sciencedaily.com/releases/2005/07/050712233245.htm

70 http://news.bbc.co.uk/2/hi/science/nature/4783199.stm

September 2006), it is reported that an area of Arctic ice melted in 2005 at a rate *18 times* the previous rate.[71]

Storms

General

There are different types of storms. Such as hurricanes, typhoons, tornados, ice storms. But there are also "storm surges". That are part of hurricanes. And these in a sense may be more dangerous than rising sea levels. Although they obviously work together. If the sea level is higher, the storm surge starts from a higher level.

And the more powerful the storm, the more powerful the storm surge.

Hurricanes

It is known that these are increasing in frequency and intensity. And can have catastrophic effects, as we see from the relatively minor event known as Katrina. A year after the hurricane only about half of the population has returned to New Orleans. And many who have not done so will never return. Further, the city of Houston that originally welcomed the evacuees now wants them to leave. There has been a big upsurge in crime.

What do you think will happen when perhaps 10 million people need to move from the Eastern and Gulf coasts of the US?

The concept of levees and pumps to protect the city forms two of the three cornerstones of the US political régime's plan to protect the coast areas of the USA from rising sea levels and storm surges. Along with the services to be provided by the Army Corps of Engineers and FEMA, the Federal Emergency Management Administration. Do I need say more about the probable efficacy of the planned treatment for the disease?

In political entities the regime is responsible for the effi-

---

71 http://news.bbc.co.uk/2/hi/science/nature/5344208.stm

cacy of treatments for diseases. The health of individuals is a function of the diseases they have. Successive American political regimes have created programs to handle the problems of American health. But in general the level of health care has gone down while the cost increases. This would say the "efficacy of their purported cure" has not been established. Yet the product—socialized medicine—has not been withdrawn from the market.

What is the chance that their purported cure for the problems associated with Global Warming will work? It is the same "Katrina team" that is working on both.

Yes, New Orleans was already in a perilous position with parts of it below sea level. But the real problem will be that of rising sea levels and greater storm surges. And many coastal areas that are now above sea level will be below the new sea level of the future.

Further, levees are not cheap. I have read the cost of bringing the New Orleans levees "up to federal standards" might be $10 billion.[72] This is half the cost of the *ten year* program of Kennedy to "put a man on the moon". What would be the cost of a levee from the Canadian Border to the US/Mexican border? How many pumping stations needed? How many FEMA personnel per square kilometer are needed to ensure efficacy of the treatment?

What if the political entity to the south—Mexico—does not want to build a levee? And will not allow the US levee to be extended into their territorial waters? Maybe a preemptive strike against them for terrorism? After all, the results could be terrifying, yes?

Again, the Internet or your local library can provide information from climatologists and other professionals in this area of knowledge.

Personally, I have found that BBC News presents the most and best articles on Global Warming of the news serv-

---

72 Peter Whoriskey and Spencer S. Hsu, *Levee Repair Costs Triple*, (Washington Post, March 31st, 2006), A01.

ices. The address is:
<http://news.bbc.co.uk/2/hi/science/nature/default.stm>

I present some articles on my own web site at:
<universal-salvage.org.uk/books/warming.php>

Typhoons

From what I read typhoons and hurricanes are essentially the same thing. When I was young it seemed that the word "hurricane" was current in the USA. But that the word "typhoon" was used in Asia.

Another train of thought says that a *typhoon* is a large and powerful tropical *cyclone* (so now we have yet another word and concept…), a low pressure area rotating counterclockwise and containing rising warm air that forms over warm water in the Western Pacific. Less powerful tropical cyclones are called Tropical Depressions and Tropical Storms. A *typhoon* is called a *hurricane* in the Atlantic Ocean, and a *cyclone* in the Indian Ocean.

But the important thing is that whatever the name, there is a very good chance they will increase in number and intensity.

Tornados

I believe, yet am not sure if this is correct, that there may not be an increase in tornado activity caused by Global Warming.[73]

I have seen a program on *National Geographic*, an organization for which I have great respect, indicating that even in the absence of a cause by Global Warming, there could sometime be a "class 6" hurricane on the Fujita scale.

If such a thing hit a major US metropolitan area it could kill thousands and cause billions of dollars in damages. And

---

73 See <http://www.tordach.orgtopics/globalchange_en.htm> This seems well documented, although I can not speak for the qualifications of those who wrote it. Which may not be bad. It is just that I do not know.

cause population relocation beyond even the scale of Katrina. And half of those seem to be not returning to New Orleans? And those remaining in Houston are now victims of "crimes against refugees..."? Something that is an expected result for a democracy.

But if this should occur it is more of a "terrible distraction" than the continuing and increasing problems related to Global Warming. Which will be more terrible and much more than a distraction. Although the "crimes against refugees" is a very worrying thing. What if there are not just tens of thousands of refugees going to a few places, but millions of them coming from thousands of places and going to other thousands of places?

Ice Storms

Global warming will almost certainly result in ice storms. In areas that have already experienced them and in areas that have not. They will increase in frequency and in severity.

They may occur all over the USA and in Europe. As a result of the loss of the warming Gulf Stream if the "ocean conveyor" shuts down. And these are not just "inconvenient" in the sense of the cold and curtailment of the ability to move about. This ice has a habit of wrecking major power lines and paralyzing a large area of society. People can not go to work and goods do not move. And with no power may not even be produced.

The 1998 ice storm in Canada was the most costly for Canada of all of their environmental disasters of all types.[74] It has been accepted that this was the result of Global Warming and its effect on the El Niño phenomenon.[75]

What happens if much of the northern US is plagued by *continuing* ice storms? While the southern part of the US—that part not inundated by sea water—experiences debilitating heat waves?

---

74 <http://canadaonline.about.com/cs/weather/p/icestorm.html>

75 http://www.sos.se/sos/publ/refereng/0003009e.htm.

Forest Fires

Rising global temperatures, desertification and drought will bring increasing forest fires. These destroy the ability of the land to hold what water there is. They add to the carbon dioxide and particulate matter in the air. And of course destroy valuable timber.

Anyone found deliberately starting a forest fire should be tried for "crimes against humanity". Which is what it is.

We have seen an increase in forest fires in Europe, the USA, Indonesia, South America. They will appear in many new places. A point can be reached at which the entire Amazon rain forest bursts into flames.

Changing Rainfall Patterns

It should be obvious that this can have a catastrophic effect on life. Witness the problems of the Sahel. Which are an acknowledged, direct result of Global Warming caused by air pollution. Mainly from the factories, power plants, and SUVs and Hummers driven by those of the North American political entity known as the USA.

Where *Dour Leader* says this is okay, and preferable to harming American industry. And especially his poll ratings when going into the 2004 presidential elections. Anything to avoid becoming a one-term ruler and becoming unemployed. Not having the *Führerflugzeug* to flit around the world overseeing political disasters.[76]

As long as we accept politics as our societal system, and use the Nürnberg Trials as a guide, it is the ruler—George W. Bush—that was and is responsible for the current results of Global Warming.

Flooding, as is now (Spring 2006) happening in Europe will increase. But so will desertification and drought. Both having profound effects on agriculture and on living in gen-

76 The German word for "leader", führer seems to have fallen into disrepute. And even in Germany they would not refer to the "leader's aircraft" as the *Führerflugzeug*.

eral. Each of them tends to create different medical and health problems.

Grazing animals do not do well in the desert. Nor do they do well when their grazing lands are under water. Especially salt water.

Among other scenarios, there will be changing rainfall patterns in the UK. Especially in the south and east. Drier summers, wetter winters. Oddly the wetter winter is bad for trees as they lose their fine roots. Then in summer—which is already drier—these roots are not there to absorb what little moisture there is. Waterlogged forests are also prone to windthrow. And various pathogens proliferate in the fluctuating water levels.

Changing Temperatures

As with everything related to Global Warming, changing temperatures can have a cascading effect. Cause drought, ice storms, crop failures. Temperatures on land—and the temperature of the land itself—will change much more rapidly than ocean temperatures. Because of the "heat sink effect" of the vast volume of the ocean. But this also helps us with the methane hydrate problem at the bottom of the oceans. However we can not dally, and must work as fast and as good as possible to keep the temperature from rising too much. Else we reach the point of no return and no recovery.

Yet a very small rise on ocean temperatures can have a profound effect on the frequency and intensity of high-category hurricanes. With their attendant storm surges. Made all the worse by a higher sea level.

Changing temperatures will bring changes in "insect life". There may be more or less insects of different varieties in your area as the highest, lowest, average and the daily temperature range changes.

With increasing night-time temperatures at the North Polar region, some insect populations no longer "winter-kill", as with the spruce bark beetle that destroyed 10,000 square kilometers of forest in the Alaskan Kenai peninsula several years ago. Similar insect problems have now cropped up with the Canadian Lodgepole pine forests which stretch from coast to coast in Canada.

Rising temperatures can mean droughts. Which will happen in the Eastern US, southern Europe and west Africa.[77]

Now, September of 2006, there are reports that animal populations and habitat are moving northwards in England as Spring is arriving earlier in the UK and all across Europe. Insects and bacteria will move with them. New insects and bacteria will replace the previous ones. With animals, the larger the size the longer a replacement will require. You can read something about this at:
< http://news.bbc.co.uk/2/hi/science/nature/5357606.stm>

General Problems

There will be profound effects on our society. Since most major port cities will be flooded. The city uninhabitable, the port as it exists now unusable.

Coastal towns and cities will be flooded. What were once beautiful resorts will be only a memory. But maybe can become exotic underwater parks? For those who will have the time and resources for recreation.

Vast areas of the US will experience continuing and severe ice storms that will paralyze much activity.

Depending on exactly what happens, places like the United Kingdom and France could become desert wastelands or be gripped by an ice age environment. Either way, it will not be like now and just survival will be difficult.

With similar drastic changes around the world people will need to move to somewhere where the weather is more tolerable. The current concept of "nation-states" (grand tribal confederacies)—already a crime against humanity—will be truly non-sustainable.

The bureaucrats associated with "immigration" (a very primitive concept) think they have a problem now. Imagine

---

77 http://news.bbc.co.uk/2/hi/science/nature/4791257.stm.

issuing passports to the majority of your citizens so they can leave. And "destination bureaucrats" needing to issue visas for a like number to go elsewhere. But then there will be the reverse traffic, from the people who feel your climate is better suited to them than theirs will be. And of course protests, riots, work stoppages and other events as individuals are not allowed to do what they wish to do.

And in a nation-state concept the arriving people would need work permits and residence permits? To get driving licenses and have their "life dossier" and its documents "upgraded".

And if neither nation-state wishes to do it then people will act as free individuals. And just do what is necessary for them to get where they want to go. So the regular bureaucrats will need to be increased by an order of magnitude. And then of course the "enforcement bureaucrats" will need the same increase. Immigration bureaucrats think they have a problem now?

In the future there may be a new class of "immigrant". Thousands at a time arriving on a cruise liner. Perhaps staging a mutiny, taking control of the liner and then beaching it at the destination.

Of course with individual freedom this extreme act would not be needed…

But what if the "enforcers" themselves opt for individual freedom? It is even possible that one of them could come up with an idea that would be a major part of the solution to our problems. Did you ever experience the "… but Daddy, why don't you do it this way?" factor? A child—or anyone who sees a new process they do not understand—may quickly and clearly see a better way to do it. As I say, "selective ignorance related to a process or knowledge may yield a remarkable new solution".

Meanwhile, there is no money because tax collections are in shambles. And no one will be able to keep the records up to date. Yes, I know, computers and surveillance cameras and things like that could do it. But they won't, as no one will conform as in the old days. As now. Because individual survival is at stake. And it would be "everyone for themselves".

"But we can't have this! This is anarchy. There must be laws and people must obey the laws. Or else".

There are two types of anarchy. The one in which it is assumed that all individuals must be controlled else there will be chaos. Our present concept, in which there is uncontrolled chaos.

The other type of "anarchy" is the "anarchy of freedom". Such as existed in the American Colonies (and later the USA) from about 1650-1850. When there was a relatively large degree of individual freedom. And the associated responsibility for the obligations and duties of individual thinking life. Often in a harsh and unfriendly environment.

And the result around 1900 was a relative "citadel of freedom" and a burgeoning economic powerhouse. But just at the point where something wonderful might be born there was an abortion. Called "politics" and "the rule of [unnatural] law".

Producing individuals were cornered. Caught in a trap bounded by the Atlantic and Pacific in east and west, and another political entity to north and south. Nowhere to go. And the non-producers—the politicians—closed in on them and began to stifle individual life. And the concept that individuals could take action and be responsible for the results. And thus maintain and build individual competence.

Which of course started to decline. Providing a feeding trough for the non-producing bureaucrats. Promising to do for the people what they could no longer do or no longer wanted to do for themselves.

Then the societal disasters of state controlled healthcare, education, pensions, and trains that did not run on time. Wars that needed to be fought because one politician did not like another. Or wanted to enlarge his slave plantation. The masculine "his" because the female of the specie had other, more moral interests in life.

Huge numbers of unproductive people. People doing

things that create no value in society. If there are 5 million political employees in the USA, then that means there is about 1×10^{11} (100 billion) person hours wasted every year. If there are more than 5 million political employees in the US then it is even worse.

There are very few political hacks that produce anything of value. The only ones who come to mind might be those of the Centers for Disease Control. And air traffic controllers. As these and all other *needed* functions could be done by individuals performing tasks required in the market. And as I said, virtually all tasks allegedly performed by politicians or those controlled by them are of no value to other thinking beings.

Summary

Massive dislocations as people seek to go elsewhere. Some even wishing to come to the areas with the greatest problems. As they are risk takers who see opportunity. But if no one could move without a bureaucrat's permission life would be paralyzed. Chaos increased.

When some of the low-lying island-countries disappear why reconstitute them elsewhere? Why not choose individual freedom instead? No need to reestablish a war department, immigration and customs, the national airline, the dog- and people-catcher organizations.

I know: too much work. Easier to cede control of your life to someone else, sit back and relax while they do their job. Then complain about the result. The problem is this has never worked. And never can work. And this time the stakes are too high.

It will not be "The Rise and Fall of [blank]". And fill in the name of a local, dearly departed civilization. The "blank" will be Homo sapiens. Who will become extinct. Slowly, painfully. In heat, cold, flooding, and drought. With people murdering each other to survive a few hours or days longer. A true Hollywood ending for Homo sapiens.

Chapter 7

Solution

Synopsis

The principle is simple. Only individual freedom will allow us to solve first the problem of Global Dumbing and then the one of Global Warming.

But while the principle is simple, the societal fallacies and their resulting distortion of the fabric of society has presented us with a truly daunting task.

No politician has been able to get the schools, healthcare or pension system working. Or technological basket cases like British Rail or the London Underground.

These—plus Global Dumbing and Global Warming—can not be fixed using force or fraud.

Preface

There is only one way to prevent or at least reduce the societal catastrophe that will result from the effects of Global Warming. And it *will* result if we do not change our societal system. The one based on coercion. That has never allowed us to solve even one of our societal problems. And it is easy. It does not involve any grand and difficult plans. Which have a habit of never working.

All that is needed is for each individual on the planet to begin assuming *total* responsibility for their own life. For the actions they take related to their life. And to do it in a way that does not damage "the fabric of society". Does not tear apart what we have now. Rather, that this non-working system will in a sense "wither away and die". As the Communists said about their system.

Instead of the "rulership of the proletariat" we will have the "rulership of Homo sapiens".

And the nice thing about this is that it can begin to put money in the pocket of all who do it. Begin to make their life slowly but increasingly better. This is done by the simple mechanism of acquiring *job security* for yourself. That thing almost everybody has been worrying so much about in recent years. But that no one seems to understand.

With job security poverty can be brought under control and ended. Rather quickly I think. Simply by understanding the true nature of job security and production compensation. With true job security comes control over your own life. The thing that has been slipping away from Americans for about a hundred years now.

This is something each individual must begin to do immediately. Why not, when it puts money in your pocket? But that is not the only reason, although a great motivator. The major reason is that in a sense we are about to be hit by an asteroid. At least we should think that way.

I have seen at least one prediction that if we do not have our emissions under control within thirty years *the effects of Global Warming will be unstoppable and irreversible.*[78] That some time soon after that we will become extinct as a specie. And conditions will be so bad that probably everything except bacteria and viruses will be gone. And many of these without a host.

Now I have also seen ten years. This greatly increases the urgency for a solution.

We *must* assume that this "worst case scenario" *will happen.* Why? Because if we assume it may not, and it does happen there will be *nothing* we can do about it. Too late to say "Gee, sorry, I will start to do something now". The penalty for not starting *now* and finding you should have is that the 3.65 billion year "experiment with life" and our 7 million year journey as the genus *Homo* will all have been for naught.

---

78 Two citations, with times varying from thirty to fifty years:
<http://www.factcheck.org/article382.html>
<http://www.washingtonpost.com/wp-dyn/content/article/2006/01/28/AR2006012801021_pf.html>

The Urgency

We must treat this situation as if planet Aristarchus were to be hit by an asteroid in, say, twenty years.[79] And we must be ready to deflect it *before* then. Knowing that if we have not changed from our society based on coercion to one based on individual freedom we *will not* survive as a specie. Not a single one of us will survive.

We will die in the most horrible agony. Physical, physiological, and mental agony. Knowing we "blew it" and there is absolutely nothing we can now do about it.

We need to have all of our free market societal functions fully operating. Have made large reductions in particulate and greenhouse gas emissions.

And have in test a prototype method of actually reversing the current Global Warming trend. If not already in use. The planting of young trees and forests that absorb CO^2 and create oxygen. I think some people refer to this as "soaking"?

You say this is "impossible"? That we can not reverse the climate trend on this planet. I do not believe that is true. The word "impossible" means that it would violate a law of nature. And it is a bit like the statement that having a computer in Roman times would have been "impossible". It was not impossible then, as it would not have violated a law of nature. The only thing that prevented it at that time is no one understood the laws of nature necessary to create a computer. Things like Ohm's Law, Maxwell's Equations, the laws of thermodynamics and quantum theory.

Within, say five years, *all* individuals on the planet must *have job security.* Meaning they have the ability to perform a number of different tasks required by the market. And are doing so. When this is true—that the individual has true job

---

79 This is the proper name for our planet, after the name of Aristarchus of Samos, one of the first great geometers, mathematicians & astronomers. The first to determine the size of our planet and of Oberth, our moon. Aristarchus also deduced that because of the relative size of our planet and moon, that our planet must revolve around the sun. A fact almost everyone forgot for about 2,000 years.

security—there will always be employment. Thus a greatly reduced chance of poverty. As in the past, a "family unit" may have a viable societal function and value.

This can be achieved by most people within one year. Five years is extreme…

But there can be no cases—*zero* cases—in which an individual is supported through resources provided as a result of coercion or fraud. We will not survive as a specie under this concept.

With credit to Beckett we can *not* wait for "we know not what".[80] We can not be like Vladimir, Estragon, Pozzo and Lucky. Unlike them, we know Godot. Godot is the *grim reaper*. And unlike in Beckett's play, if we wait for Godot we will meet Godot. And for the brief moment we exist after the meeting, we will wish we could change things. But we will not be able to do so.

Background

This solution chapter will be brief, containing mostly principles. It is necessary and expected that individuals will begin to think and to take control of their own lives. Those that can not may and should fall by the wayside. It is the way of nature, of the natural world. Our current attempts to contravene that have consistently failed. And are a major part of the reason we are now in "extremis" as a society and specie.

When I made a comment similar to the above to someone for whom I cared very deeply, she became very angry. And accused me of being "mean" and all sorts of similar things. Of not caring for my fellow man. In the sense of the specie. But when you think about this it is not true. It is an act of kindness to tell someone they are not doing something correctly when this is true. All the more kind if you can also tell them the reason and the correct way of doing something.

---

80 Samuel Beckett (1906-1989). Irish playwright (although he might say he was French), and powerful and enduring figure of the theatre of the absurd. Appropriate for politics, yes?

It gives them a chance to learn and to improve their life.

Telling someone they are right when they are wrong harms the person. It is not an act of kindness to make them temporarily feel good about themselves when their thought or action is wrong. They can not improve themselves. There are many things wrong with this action. I thought to describe them, but it would be beneficial to you the reader if you think of and about them yourself.

Having said the "bad" thing in the first paragraph, almost all individuals are capable of thinking and taking control of their own lives. And have the *ability* to learn necessary new things to accomplish this if they have the opportunity. They must now also have the *desire* to do that. The incentives are great.

Life for Homo sapiens need not be all that much different in the future than now. The three crucial things that *must* occur:

1) There must be total and complete individual freedom for each and every individual on the planet.
2) Each individual must have job security.
3) Each individual must stay within their allowable "carbon footprint".

There can be no "technical" doubt about what is happening. The twin problems of Global Dimming and Global Warming. Made all the worse by Global Dumbing. A phenomenon created by politics and politicians. Who seek to do everything for everyone, thus depriving their victims of the chance of making mistakes and learning.

For Global Warming, the only open questions are how long it will take to reach a given temperature. The ocean temperature required for the methane hydrate layer to start releasing methane. The same for the permafrost and "methane lakes" in Russia. Although these lakes are relatively less deadly, it is cumulative and we do not want it to happen. And whether we can reduce particulate and greenhouse gas emissions soon enough to give us a "working margin" to prevent these things.

What will be the likely rise in sea level and when? With this we can plan ahead to avoid the worst of the catastrophes that will result from rising sea levels.

As I say, it is best to consult the works of accredited climatologists for information on that subject. The only thing I really know—and know better than anyone else on the planet—is the cause of our societal problems. And the principle for the cure. Including the societal problems resulting from global dimming and warming.

If we can not take the appropriate steps, the 3.65 billion year "experiment with life" on this planet in this star system *will end.* Not to be resurrected as there is probably not time before the sun starts its own "dimming". Before the fire starts to go out. Even assuming that Aristarchus completes the requisite geological and biological cycle again. Which may be unlikely.

Further, within about 100 million years this planet will be totally unrecognizable to those familiar with its appearance now. Continents as we know them now will be gone. And of course all their cities. Which means that in a few million years or less in the future we need to have made plans to move elsewhere in the universe. Or to have learned how to prevent what will naturally happen from happening. Which we could conceivably do if have the technology to do it.

I find it all incredibly depressing. To think we came so close. Certainly if we had understood a hundred years ago, when we really started using fossil fuels "as if there were no tomorrow". No day of reckoning. Even twenty-five years ago if we had understood the significance of Dr. Stanhill's 1980s data related to his readings in the 1950s.[81] Yet there were other crucial inputs not yet available or understood.

And of course if we had not adopted a societal system based on coercion and continued to use it. Did not have

---

81 Dr. Gerald Stanhill, Ph.D (Emeritus). The Volani Center, Israel. Probably *the* original source for the data that began to confirm the concept of Global Dimming.

"government without end" as I understood in 1956.[82] About the same time Dr. Stanhill was doing his original work.

I am not a climatologist, but I think they might say that technically it is certainly possible that we can fix our problems. But that as a collection of 6.4 billon slaves to thousands of political rulers we simply can not do it. We must be as nature intended, individual thinking beings.

And this is the *only* way we may do it: *as individual thinking beings*. Each of us responsible for rationally and morally discharging the duties and responsibilities of individual thinking life.

But our chosen and maintained social system based on coercion has not allowed us to solve even one of our societal problems. And with that system we will not be able to solve our current pollution problems.

Think of the *minor* disaster of Katrina. Minor compared to 10 million people needing to move from the coastal areas ranging from Maine to Texas. The move was very difficult, very expensive and not very successful. And at first Houston welcomed them with "open arms" and "open pocketbooks". But now, a year later, they want them to leave. As they have produced a lot of problems.

There were floods in Carlisle, England in January 2005. The dislocation involved hundreds of people, not thousands, tens of thousands or millions. A year and half later, from flooding that was very minor relative to what will happen in the future, there are still families waiting to get back into their homes. And the cost was less than $1 billion. This is more than I earn in a week, but still much, much less than the cost of the Katrina problems. And will be minuscule compared to the costs of relocating 100 million people.

---

82 There is a chapter on this subject in my book *Cowardly New World*, (Universal Salvage Corporation, Escondido, CA, Aug 2006). And you can read an essay on the Internet at:
<http://www.aristarchus.org/nyt_gov_without_end.php>.
You can order the book on-line at:
<http://universal-salvage.org.uk/book_list.php#cnw>

If there was total individual freedom and self-responsibility for action, individuals would simply relocate themselves. Simple. And it does not disrupt society.

To solve our problems we must very gently and *carefully* disengage from the deadly embrace of politics. All three branches: civil, military and religious. They are like an anchor, dragging us to our death.

You will not be alone. Everyone on the planet must in principle go through the same thing. There will be friends and colleagues with whom you can consult. To see what has worked best for them. Or to tell them what has worked best for you. It will be a little bit like the time from about 1650-1850 in what became the USA. The individual freedom, responsibility for life, and working together that created what would become the economic powerhouse of the world. It was individual freedom that created that. If you want a little information and inspiration read *The Mainspring of Human Progress* by Henry Grady Weaver. Almost ironically, he worked for the basket case of American industry, General Motors. None of them read his book. Or if they did, they did not understand it.

The process will be very much aided by the vastly improved communications of today. Especially the Internet.

If Mr. Weaver's management had read his book and understood its message they might now be the only motor car manufacturer on the planet. As no others would have been able to withstand the competition. And as some have alluded when comparing the automobile business with the computer business, a Chevrolet might now cost $1,000 and get 1,000 miles per gallon.

Although this might not seem to fit here, please remember that all knowledge is related. And thinking about individual freedom and the problems of General Motors I thought of the Sahel and their problems. And you probably think there is little connection?

With complete individual freedom and responsibility for life—meaning no despotic or even benevolent rulers—all those of the Sahel would have had job security. Thus not be living in poverty. They would have quickly seen the disappearance of the monsoon rains. And moved elsewhere. Or changed their local environment and life. With complete individual freedom they could have moved to anywhere on the planet.

Further, there might have been one among all those highly educated and literate people (that goes with individual freedom) who, having a very broad knowledge of physics, might have told us what was happening and why. And what to do about it.

Continuing an interrupted thought, the "word of mouth" network that existed during periods of individual freedom and greatness in America has a modern counterpart: the Internet. This would help you to do the things in life that would make your life slowly and continuously better. As long as you assume responsibility for that. And do not cede this to others.

Value in life and society

General

Think of the things you like to do in life. The things you want to do in life. The things that add value—add enjoyment—to your life.

Why should you do—be required to do—things that do not add value to your life? Very few people are truly insane. Wanting things that harm themselves or others. There is also the thought that we usually do best in those things that truly bring us pleasure.

An individual becomes the world's greatest painter, bicyclist or climatologist because it is something that really interests them. Something they feel driven to do. And you have someone who adds quality not only to their own life but to that of others.

As I say in my essay on *Value in Society*, each individual thinking being must ask themselves from where it is they receive value in society.[83] To encourage, value and patronize those individuals who provide these things. And to not patronize those who do not provide things of value.

Why mindlessly follow those who encourage and create wars and acts of property destruction? Those who seek to make you pay more for things than you should? In most political entities if an individual did that they would be persecuted by the political state. Yet when the political state does it, it is legal. Not only that, to do otherwise is illegal and a crime. So perverse is the concept of politics that it is often illegal to do that which is moral and rational.

Or not be able to trade with others simply because the ruler or his agent does not like them? Or because they are from a different tribal entity (country) or religious group?

Think. Why do you need a license from your ruler to be married? Why must you pay a bribe to your ruler to "import" a product? An import fee or "duty" is nothing more than a bribe to the ruler and ruling party faction. Why must you receive permission from your ruler to travel to another area? Or to be allowed to enter another geographic area of the planet?

How does regulation of your life enrich your life? Your ruler wishes to tax something coming into your tribal area. Perhaps to "protect" the capability for making something he may need if he wishes to go to war. This makes life more difficult for those in the other political entity. And their products more expensive for you.

And if he does decide to go to war—perhaps because someone once threatened his daddy—you will be required to go to war and "kill or be killed". On top of that, you must pay for the war. With a "time payment scheme" that may take a decade or more to pay for it.

So other rulers tax your products. And now the victims—usually called citizen—of each ruler again have their quality of life reduced. Where is the value? Only for the smiling faces of

---

83 You can read this in my book *Cowardly New World*, or on the Internet at: <www.aristarchus.org/value.php>.

the "photo-op" who are talking about how to control your life over a sumptuous "working dinner" with the best wines. And their freedom from traffic jams to preserve their security from the adoring masses.

Except for those who do not bother to protest, and pay five times what they should for a T-shirt and another 20% for gasoline or petrol for their car. And are told this is the price for freedom and security.

You really need to think from where it is you receive value in life. And use only those services. So that the providers of things that do not bring value to your life find they have no customers. And stop providing the non-services for your life.

It does not require attacking anyone or interfering in their life. It only requires getting on with your own life in a rational and moral manner. And assuming responsibility for the actions you do take. Learning from them to increase your *own value* in the marketplace. Which will be synonymous with and increase your own quality of life.

If you cede or have ceded this control to others in the past, then you have only yourself to blame for the results. If your political agent does not provide you with what you want in life then you should consider trying to provide for yourself. And since no one individual can provide for everything they need in life, this will provide tasks for others to do. The summation of which is "a job" and better life for all.

Do not do those things that create "immorality waves in society".[84] Things that are immoral. In the long term this will have a negative effect on the quality of your own life. Your credit rating will be effected…

I will tell you some things in general that must be done. I have detailed thoughts on many of them that are probably valid and workable. But I will provide those in another book.

---

84 For more on this subject read my book *An "Einstein" Thought Experiment: Rationality and Morality Waves in Nature*, Universal Salvage Corporation, Escondido, CA 2006).

Or two.

Or I may not. It depends on Homo sapiens. I am old and—because of primitive biology—I will die soon. If no one seems interested in doing what is necessary I may simply enjoy my last days with my young wife and very young children. It is up to you…

Ask me to keep working, please. Most people do not understand. But in a sense "play" and "work" are reversed for me. I like to work. It keeps my mind active and, hopefully, healthy. Which keeps the body—the life support mechanism for the brain—healthy. Stop thinking and your brain atrophies. Then your body. For me, the things most people look on as "play" are work. I become bored, just being a spectator in life. Unless they are my children, then having time to be a spectator is very nice.

If you want to hear more about the solution to our problems, you may send me a message through the form at:
<universal-salvage.org.uk/msg.php>.

Human potential

Most people can learn new things and learn to do many things. They only need to believe this is true and try. If you believe that you can not learn and do new things then this will be true for you.

Solution Concepts

Background

It is very difficult to know what to write. For two reasons. One is that the principle is simple. But on the other side the millennia of societal fallacies make it very difficult for those cherishing their fallacies. To understand the natural condition of thinking beings.

A condition that existed in the American Colonies and the USA from about 1650-1850. And laid the foundation for the eventual economic power of the US. And it is the ending of this concept of individual freedom starting in about the

1850s that started the decline of Homo sapiens in the US and the rest of the world. To the point where we are now a very endangered specie.

Americans are in worse danger and condition than were the Russians of the former USSR. Americans believe they are free when they are not. The Russians suddenly became freer if not free, and were afraid of the concept of individual freedom and responsibility for the duties and obligations of individual thinking life. Under the system of the USSR they *knew* they were not free. They had no illusions on the subject. Because of this, may of them learned to take care of themselves in the "underground economy". But in this respect, the control is much more rigid in America than it was in the CCCP. And Americans are much more puritanical and willing to accept their servitude than individual Russians were.

I think Americans—long believing themselves to be free—do not understand enough about the situation to be afraid. And, sadly, do not really want to undergo the rigors and responsibilities associated with individual thinking life. They want someone to do that for them. And then complain about the results.

Each individual must assume responsibility for their own individual life. Maybe I seem to use the word "individual" too often. Repeating it, stressing it. But it is extraordinarily important. To assume responsibility for the results of your own individual life.

For its successes and mistakes. Which will increase individual competence and make life slowly, progressively better for each individual. And the summation of individual life is societal life. Specie life and survival.

With our societal system we have abdicated first responsibility and then control for our lives to others. To the political agent representing the political state. And we spend more time nitpicking and interfering in the lives of others and less time controlling our own lives. Partly because with our societal system we often have little or no control over our own life. And vent our frustrations in telling others what to do with their lives.

Motivation

To understand the problem you should read the Horizon program narrative. Available at: <http://www.bbc.co.uk/sn/tvradio/programmes/horizon/dimming_trans.shtml>.

For inspiration, read *The Mainspring of Human Progress* by Henry Grady Weaver. This shows what made America the most powerful political entity on the planet: individual freedom. It is a nice story of individual achievement. And a great deal of very good and interesting US and world history.

As individuals we need to become the same as the "frontiersmen" of about 1650-1850 in the US. To be individuals in charge of our own lives. We must return our societal system to that model. No bureaucrats or "regulators" to tell individuals how to conduct their lives.

With true job security and a proper understanding of production compensation you build you very own, personal "safety net" for your life. You create a "level playing field" for yourself and allow the same to others.

Basics

As I said in an earlier chapter, individuals must reclaim control of their own lives. I will repeat some of this because it is very important.

Individuals must observe the following principles:

1) *Each individual may do anything they wish that results in moral action.*

   This is the natural condition of a thinking being.
2) *Each individual—and only that individual—is responsible for the result of their actions. The rationality and morality content of their actions.*
3) *Each individual must have skills required in the marketplace that may be sold at a discount so they and others may make a profit on the sale of those skills.*

We need a few other things:

1) A "not to exceed" level of particulate and greenhouse gas emissions. And a way to measure it. Our individual, allowable "carbon footprint".
   This number multiplied by 6.4×10^9 will also provide the level for Homo sapiens. Or the reverse: take the total allowable emissions and divide by the population. This is the "not to exceed total" for the individual. The output of some producers may be understandably and excusably higher. This means some individuals must be lower.
   Individuals will need to work with companies and entrepreneurs to ensure that as a group they do not exceed the index.
2) Clearing houses that will maintain the "pollution index" for each individual.
   This could be a "credit clearing house" that would also have the credit rating for each individual. And other things, such as their "soaking index"? Which I will mention more about later. The crucial part of the "regulatory" mechanism of society. Refer to the earlier chapter as a refresher.

As an individual you must create job security for yourself.

General

What about the "Dodo Birds" of our current world? Functions such as immigration and Customs. Marriage and business license offices. Mayors and supervisors. Senators and representatives—both from the "federal" and "state" sectors. The war, state, treasury and police functions. Such as the SS, the euphemistically-named "Homeland Security", Marshals, FBI, CIA, NSA, etc. Prime ministers, and ministers not in their prime. "Front benchers" and "back benchers". Those on the left and right side of "the aisle". In England, those from the "Home Office" and the "Colonial Office". I think they no longer use the name, but whatever part of the régime has the same function, controlling the effective colonies of Northern Ireland, Scotland and Wales. A few of these will be discussed in slightly more detail following.

In the European Onion [sic[, the need to expand.[85] From the current 25 nations to something more like 400 million sovereign entities. At which point there will be individual freedom. And I do not think that any of these need to have their own army, navy, air force, national airline, central bank (a banking monopoly), immigration, agriculture department, health and education departments, etc. The relevant individuals will most likely be busy making a better life for themselves.

Think of the money this will save! The entire budget of the European Onion. Which comes out of each and every victim of the system. And real estate it will free up for moral and rational uses. The tremendous reduction in the "collateral damage" caused by their activities. The streets no longer blocked for receiving the stream of unimportant people. The politicians of the world.

City fire departments are ideally placed to take fees from their subscribers. Such as Rural/Metro Corporation does in Scottsdale, Arizona. It is a private company offering fire protection services. For those who wish to have fire services, they can subscribe.[86] This is also true of the park services and forest rangers. These are some of the very few existing organizations that have a reason to exist.

Another service that should continue is that provided by the Centers for Disease Control (CDC). With colder and hotter temperatures will come additional and new medical problems. This could take the form of an "association". With business supporters and individual subscribers.

Money

What about money? One of the best monies disappeared

---

85 Well, it is coming unpeeled, yes? Is much more like an onion than a union. And it has other attributes of an onion.

86 In the case of Rural/Metro, I know at one time their protection services were so much better than that of the city fire department that the reduction in insurance costs was greater than their service fee.

decades ago. The "ride coupons" from the original Disneyland in Anaheim, California. As I remember, the A, B, C, D, and E coupons. Another form of money would be the vouchers issued by barter organizations. Common stock and mutual funds. Mutual fund shares especially. There could even be an agreement that existing US Dollars and Euros have value. Providing no more are printed. No need to create inflation just as we are starting a system that works.

Permits and Licenses

Marriages and divorces

I don't know who ever had the bizarre idea that the political state should be involved in marriage and divorce. I suppose it originated a long time ago, in tribal antiquity. But definitely has no place in the life of a thinking specie. Marriage is a personal concept, not a political concept.

If two people say they are married to each other, then they are married. If one of them says they are not married, then they are not married. And no political law will change the truth and reality of that. Whether it is the civil or religious branch of politics.

So they should all just close up and go home. And none of the records are needed. If individuals want to keep a record for historical purposes that is of course their business. The rest of the papers can be recycled as indicated elsewhere.

If two people who previously said they were married now say they are not—one of them says that—then they are not married. And they should sort out any problems resulting from the change. Which should actually have been agreed to *before* they entered into marriage. Then there would be no problems now.

This is one of the major problems associated with the current concept of marriage. That of only needing the approval of the political state. People then leave it to the state to sort out the problem if they change their mind about being married. And then usually do not like the results.. So, as I say best to sort out the situation before getting married.

This would also free up a lot of judges to do something worthwhile with their life. Such as judging hog-calling contests, beauty contests, and other moral and rational things.

Anyway, that gets rid of a lot of unneeded bureaucrats. And an office full in every village and town in the world. And think of the paper saved! Many lawyers freed up to do something worthwhile also. And all their reference books can be recycled to make paper and save trees.

Professions of all types

Bureaucratic licenses are of no value and are not needed. An individual is licensed by their reputation in the market. Or perhaps by an association or what would now be called a "union". For medical doctors in the USA their union is the American Medical Association. Individuals might also be "licensed" (recognized) by an insurance company. The lower their insurance premium the higher their license rating. Or by their reputation in something such as the US magazine *Consumer Reports* or the British magazine *Which?* I suspect that most geographic areas of the world have their equivalent magazines or publications.

Driving

An ideal organization to "license" a vehicle driver is an insurance company. They might have "grades" of licenses, perhaps by the premium paid. A high premium indicates a risky individual? Or the individual and company that owns road and street systems. Authorization from them to use their roads and streets would serve as a license. It would not be good for the road and street owner if they let people who caused accidents use their system.

And of course the preference of an individual for certain road and street companies represents a "market endorsement" for these companies. So it works both ways. Like recommending a dentist or auto mechanic. Road and street owners who maintained their properties in the best possible condition and had innovative ideas that improved their utility and costs effectiveness could receive the highest possible market endorsement: that of becoming a moral and rational *monopoly*.

Lawyers, Prostitutes & Non-Elected Bureaucrats

Lawyers

The world's oldest and most disreputable profession has a chance to re-invent itself. And provide worthwhile services to Homo sapiens.

In a society based in individual freedom there might be a need for "contract engineers". Individuals to broker deals between producers and entrepreneurs. People or innovators who perform tasks that others re-sell in the market. The latter—entrepreneurs—operating very much like a "company" or "corporation" does today. Or more better a "holding company". The big difference being that the "company" has no employees. Only "Production Associates".[87] With the "company" effectively outsourcing all of its tasks.

The best Production Associates might have an "agent", as film stars, musicians and sports figures do now. And as lawyers do now, they could become agents.

Prostitutes

Usually meaning a sexual prostitute. A professional or amateur who sells their own sexual services. One of the world's older professions. And one which would be expected to slowly disappear. As individuals acquire job security in a different market area performing different tasks.

In the absence of force or fraud, prostitution is moral. It may not always be rational, but that is for the account of the individual professional. And not a matter on which producing individuals should waste their time or effort thinking about. Or, worse, trying to take any immoral actions. Such as interfering in the life of the prostitute.

---

87 This is the subject of a forthcoming book, *Productivity Galore*. To be released later this year (2006) or early next year (2007).

Non-elected bureaucrats

As Thomas Paine might have said, "Of more worth to mankind is a single trash collector ["Sanitary Engineer…"?] than all the bureaucrats who ever lived".

What to do? Some bureaucrats and politicians are well-meaning and kind people. But just in the wrong or in no profession. However for some period of time, as discussed under "Regime Salaries" in this chapter, their salaries should continue at a diminished level for a time. To allow them to make a not-too-painful transition to performing useful tasks in the market. Which they should do with all possible haste. Some may have some very useful skills that would facilitate this. Some may have long harbored the idea and wish to "start a business". Or otherwise "strike out on their own". Now is the time.

Housing and Living

General

I remember some years ago reading something in a UK newspaper. The régime was agonizing over "where will the money come from for 'x' million new housing units that will be needed in the coming years?" I thought to myself: why not from those who need them? If they need an apartment they rent one. If they need a house they buy one. Finished. Market demand will ensure they are available.

Although this is not the same thing, the principle is the same: someone in Spain admonished me for turning on too many lights, not turning one off, or wanting a new one some place. Don't remember which. "We are a poor country. We do not have much electricity." I suggested building more power plants. "We are a poor country. We can not pay for the gas, oil, coal or whatever to fuel the power plants." I said "The customers who use the electricity will pay the power company and they can buy the fuel." "But you don't understand, we are a poor country…" She was correct: I did not understand.

The principle is that individuals create a company to produce a product or create a service in response to perceived market demand. If they are right they will make money. And can purchase raw materials, pay employees, make a profit and expand their business. If they are wrong they have a problem they need to solve as individuals.

As I say about a principle of capitalism: *In true capitalism there is no guarantee of success. Only the guarantee of the full and unhindered right to try for it.*

After that digression, continuing with the idea of cities: in a free world cities could be an efficient concept. Living in close proximity to work with its reduced travel. And having many services available, including entertainment. The availability of knowledge centers.

But there must be some changes. The name may still apply to the geographical area encompassed by the *ancienne régime*. But the régime "withers away" and associated bureaucrats find something productive and profitable to do.[88]

I will not presume to tell you in detail how to handle the roads and streets. But in principle it is no different than a telephone company or railroad. There are "lines" and "rails" to be maintained. The streets and roads. Many different companies might own and operate the roads. As with an intercontinental telephone call, a car may pass through many companies. We have long had the technology to automatically collect a fee. And the road companies that provided the best roads would have the most customers.

Charging a large fee might reduce traffic and congestion. But too large and you might lose more customers than you could afford to lose and still make a profit. Bad road or street and your customers would choose another route and vendor. No different than any other business.

---

88 Remember Galambos' definition of *profit*: "Any moral increase in happiness".

Improving our houses

We need to make them more energy efficient. And can do it relatively cheaply and cost-effectively. Add or improver insulation. If you have an attic, add a reflective film to the inside of the roof. If your roof is a dark color, paint it white or a light color. Add or improve weather stripping around doors and windows. Drapes or curtains on windows facing the sun. Energy-efficient lighting. Mini-fluorescent bulbs. Low voltage, long-life bulbs.

If you live in the country you can make a natural air-conditioning system. As the Romans did. Visit:
<http://www.thefarm.org/charities/i4at/aircond.htm>.

See also:
<http://www.eattheasy.com/live_naturalcooling.htm>.

You can sleep better in hot and cold weather with a wool mattress or "mattresstopper". Wool naturally wicks away moisture in hot climates. Plus it is great for cold weather. And can provide a protective insulating layer. With a latex covering for the mattresstopper or for wool pillows it is hypoallergenic and you will have no dust mites. It is helpful for those who have asthma, rheumatism and arthritis. Reduce both heating and air-conditioning bills, and emissions.

Miscellaneous

Controlling the air waves

Little different than with the property known as "land". Except air waves do have an advantage. In a land area there are only so many street corners available. Yet the bureaucrats have passed up the opportunity to control street corners. Except of course under rigid Communism.

But with airwaves, technology can create more "street corners". More usable bands. When I was a kid 2-meter VHF at about 144 MHz was nearing the usable limit. Certainly UHF at 288 MHz was really exotic, "high tech" stuff. Now we use the 50 GHz band and higher.

No reason why individuals should not buy, own, operate,

and sell frequencies. As I have said, in a free society there is never a danger from a monopoly. Or from someone who puts such a high price on something that no one wishes to buy or use it.

Think about it…

Production and living villages

Something like this could be "environmentally friendly" also. And actually nice for living. Sort of a "cluster economy". In the right area and with imagination and innovation they might be energy independent. With hydroelectric power, wind power, fuel cells, solar cells and something not yet discovered.

With each individual being a "personal fabrication center". Using the so-called "FAB technology". In a sense an individual can adapt the "FAB" concept to make almost anything you can imagine. Or to fabricate the very common and mundane things desired in life. A person can truly be their own company.

During World War Two the Japanese used this concept in principle. Householders would fabricate the individual pieces of very complex machines. With some having the job of assembling smaller bits into something larger. And periodically vehicles would come along to collect the items produced to take them to the next stage of production or for shipment to the customers, and to pay for tasks or jobs completed.

Innovative Technical and Business Actions

Background

As we all know, there are many, many things that can be done. Many known now, but most of which we will discover only as we start to think about what we should do to handle Global Warming.

Much of what I have said here will initially appear to be irreverent, traitorous, blasphemous, heretical, etc. But if after some thinking it continues to seem so, then you have missed the message. It is rational and necessary for our survival as a specie.

Individuals can monitor the amount and type of energy used. And its “emission quotient” or “carbon footprint”. And take steps to be below—way below—current guidelines. It will be profitable for them.

It has been shown that using recycled paper saves an impressive amount on emissions.[89] And there is a *huge* amount of that available. All of the papers created and/or kept by political entities can be recycled. There is obviously no need for it in the future. Not even as a memorial to our barbaric, primitive past.

Store CO^2 underground or in the ocean

Individuals are working on this idea now.[90] Not only could it lower atmospheric CO^2 levels, but it would be a profit-making business opportunity.

Grow trees

This has been shown to create oxygen and remove CO^2 from the air. Especially with young trees. And when they are not young, then they can be used for something else and the cycle repeated. I think this is known as “soaking”?

Oil from plankton

A Spanish research team is working on this.[91] It could become a reality. Phytoplankton absorb CO^2 as they grow. “Farming” them could reduce CO^2. It is said that the fuel thus created is free of sulfur dioxide and would be cheaper than the current price of oil.

---

89 Read about it at: <universal-salvage.org.uk/pdf/RecycledPapersCutsCO2.pdf>

90 Read about it at: <universal-salvage.org.uk/pdf/JapanPumpCO2Underground.pdf>

91 Read about it as: <universal-salvage.org.uk/pdf/SpanishMakeOilFromPlankton.pdf>

Other

I am sure there are many other processes and technologies currently available to reduce emissions. And for alternate fuel sources. And know that the future could bring startling new ideas and technologies. If nothing else, "necessity is the mother of invention"? And the opportunity for profits may be its father?

Think of a cure for AIDS. If the individual creating it received $1 for each injection. I think I might be happy with 10¢. That might be $600 million for the first round of shots.

With that much money the first year and perhaps even the second year, I might create a research lab that would create a fuel cell that would allow electricity to be produced at less than 1¢ per KWH. Or find a way to make Kepler [Mars] habitable so that people tired of the problems here could move there. And think of the money I would make on the "land boom"? I am currently selling some nice plots on the Kepler River...

Some Specifics

General

I can not cover all functions, operations and activities needed in a society of thinking beings. But will cover only what I hope might be a representative sample. Again, individuals *must* think for *themselves* as individuals and in the long term. Individual self-interest has a bad image. But it should not when individuals think in the long term. If they think what is best for themselves in the long term they will not take actions that intentionally and adversely interfere with the lives of others. As this may cause retribution. And waste time and resources for both individuals. Time and resources that could better be used in doing those things that actually improve individual life.

It is not in the best interest of the individual to do anything that will interfere in the life of another person. This causes "local" chaos and disorder. And when multiplied by a large number of people is societal chaos and disorder.

Individuals should *not* directly challenge their local political state and rulers. They should have their tacit assent for any action. Remember, they do control the political police, political "justice" system and political "correction system" (prisons). And in anything where they do not agree they will be the "winner". Although this is only in the short term, we all must also live and survive in the short term. Which has a habit of turning into the long term.

Something I want to say here, and which I think I will say again at the end. As it is very, very important.

There are many people today who have enormous financial resources at their disposal. How they achieved this position is not important for this discussion. But what is important is how they choose to use these resources between now and the time we have ensured our survival as a specie.

Without that survival, there is nothing. So I would ask that these people either use these resources to assist in our survival or not use them at all. They will truly have a Pyrrhic victory—including the heat and flames—if they do anything that degrades our ability to achieve survival of the specie. Or even if they do nothing to help when they could.

Actually, there would not even be a Pyrrhic victory, as Pyrrhus did defeat the Romans. Just at a terrible cost. But if anyone tries to derail our attempt to solve the problems of Global Dumbing and Global Dimming they run the very real risk of accumulating vast treasures of no value. As *no one* would survive to enjoy them. Not with "all the money on the planet".

After survival is ensured, if they wish to return to the game of politics—doing things that are immoral and/or irrational—then they should do as they wish. But please, not until our survival as a viable specie is ensured. By making the right investment now a person could achieve immense and enduring importance in history. Have the gratitude of all mankind. And vast treasures besides.

What might a royalty of, say, 0.01% of the gross planetary

product in perpetuity be worth…? I think it could be on the order of 6 x 10^8 dollars the first year, Increasing at, say, 10% per year for a few million years.

And when we have life spans measured in millions of years? Which *will* happen in the future if we survive this crisis. That would keep one out of poverty for some time.

Regime salaries

For all functions understand that salaries, probably at a reduced level, should be continued for a short period of time. A bit like "unemployment compensation". This means that a portion of taxes should be paid. As if you are "paying for services rendered".

The debt of all political régimes is something that will never be repaid. It is best to simply forget about that. But some régimes may have a lot of cash on hand. With more trickling in. And there is no need to fritter it away (if one can be so gentle) on such things as a war in Iraq. Right there, and by dissolving the war machine there should be trillions of dollars suddenly available for useful and reasonably moral things.

With just the salary of the chief warlord, "Rummy," you could probably seed several good businesses. The battle tanks that he controls have some scrap value. And since they belong to "all the people" they could be sold for scrap. Unless that idea of "ownership" the régime has told us about is false? The steel could save a lot of energy and emissions by being recycled. They might also be usefully employed as artificial coral reefs.

I am certainly no expert on this, but part of the atomic weapons might be converted to fuel for nuclear power reactors? Although we should be cutting back on power usage, more of these may be required in the future.

The "common" war explosives and ammunition might be dumped in a volcano? Well, we do need to use our imagination in getting "the most bang for the buck" from this conversion to individual freedom and responsibility for life…

Customs and immigration

Customs is a perfect example of an unneeded "service". Something that increases the cost for both producers and consumers. And then as a final insult to rationality incurs a huge "brokerage" charge. The cost of maintaining the useless organization. The people and their offices and equipment. The immigration computers of the US régime buried deep under ground in San Diego and other places should be converted to doing work on climate research.

Immigration provides the least needed of all non-services previously supplied. One that is actually very detrimental to solving the societal problems arising from Global Warming. These functions should be stopped immediately by all political regimes. It will free up a huge pool of manpower of both sexes. And a lot of money that individuals can keep for themselves.

With huge savings to producers and consumers. All Customs and immigration offices should turn off the lights and lock the doors on some Friday. There need be no problems with immigration. But again, this *must* be done on an agreed, "consensus" basis to avoid societal disorder.

Normal travelers come prepared for a period of stay. They have money, clothes, something to do. So they are no problem. Other travelers might have a problem. They must think what will happen to them in a strange land with no money. If they have friends or others who will *voluntarily* (without force or fraud) help them, fine. If not, they need to find a job soon or they have very major problems for themselves as individuals to handle. They may in effect have committed suicide by bad judgment. As those who arrived without money in the American colonies from the 1600s to the early 1900s might have done. There was no "safety net" for them.

A few will become very rich. A few will die. The majority will somehow get along. The old normal distribution or "bell curve". And word will get around that it is not easy and perhaps best to chance it at home and try to make something work there. And a much better chance when there is no political régime to interfere in their life. Simple, yes? Simplistic? No. It is the way the world should work and will work.

We all know it has not worked with the current system.

Farm and other subsidies

In the 1940s when I was a teenager in school and working afternoons, evenings, and weekends (without a permit) I was told that some farmers were being paid to raise or plant things. And then being paid not to raise or plant different things.

From my earlier, elementary school education, I knew this was crazy and at first did not believe it. I could not believe that any organization would make such payments. Nor could I believe that a rationally thinking being would accept such payments. As it was basically immoral.

Obviously I had a lot to learn about "adult" life…

This is something else that is simple to stop. Just stop it. The recipients then need to do what any other organization intending to make a profit would do. In the long term it is irrational, bad for self-esteem and for individual competence to accept handouts. It also greatly "muddies the economic waters".

Find a product that has a market demand. And at a price you can produce it and make a profit.

Otherwise, to have the hallowed "level playing field", simply pay every organization producing anything a subsidy. Computer producers, TV and stereo producers, travel companies, pizza restaurants, prostitutes and anyone else producing anything. Why not? It makes as much sense.

Or maybe the farmer is in the wrong business? Maybe they should grow fish or shrimp? Or trees and sell carbon credits? Raise bats and sell bat guano. Something other than what they are doing now.

Social security retirement

For perhaps the life of most retirees, this can be funded

from the money that would have gone to the war department and its war lords. And from the sale of scrap such as tanks, war planes, ships, etc. Most people should not have decided to stop working and wait to die. Many of them have useful skills that would earn money for them.

There should be no further confiscation of assets to fund this illicit scam. As discussed elsewhere individuals may and should create and fund their own "retirement" program if they wish to stop working at some time in the future. For, as Galambos said "… no one should be surprised they have grown old".

It makes about as much sense to stop working at 65 or some other arbitrary age as it does to eat simply because it is 5:00 pm. You should eat when you are hungry. And if you are not hungry, don't eat. Stop working if you do not wish to work more. Or work less at your regular occupation and do some other things you might like. Since you like them and may be good at them, then you may also make money at them. With a rational system of production compensation this is very easy to handle.

The only thing the American War Department—euphemistically referred to as the "Department of Defense"—is defending people from is reality. I suggest this is not a valid product for a thinking specie.

The Americans have not had to defend themselves for at least fifty years. During that time they have only attacked other countries. Handily defeating Grenada and Panama. But having a problem with Vietnam, Afghanistan and Iraq.

Most individuals should never have retired. Many of them have very useful skills. The employment of which was arbitrarily cut short because they had reached a certain age. Retirement and death are synonymous, with both to be avoided as long as possible.

It is crazy. When a person should be one of the most skilled and valuable individuals in their profession they say—or are told—that they are now useless. They should stop thinking and prepare to die. "Take your watch and go home".

This is barbaric.

If people want to work less physically, okay. But they may also want to do more thinking. And it is this that creates new and useful knowledge and products. And keeps the physical body in good shape. Adds years to your life and probably puts money in the pocket.

Zoning commissions

Another huge boondoggle to get rid of. And this one with a lot of paper. And no worthwhile service. So much so it is difficult to think of anything more to say…

Education

Only teachers—real teachers—need work in education. No administrators required. Any competent teacher can act as their own administrator since they will have very little of it to do. A competent student or two could handle most such chores. And learn some valuable additional lessons. And all good teachers know that the teacher always learns more than the student. So here is a great opportunity: start teaching.

It may be a young science student with an uncluttered mind that makes the innovation that literally saves Homo sapiens from extinction.

Such as a William Thomson (1824-1907) who wrote so many dazzling papers as a teenager that he issued them anonymously to avoid embarrassing his professors.[92]

There must be an emphasis on science and technology, positive history, laws of nature and "how to know if you are right". The use of the Scientific Method. Remembering that "science" is merely organized, useful knowledge.

An emphasis on how to think and why things exists or are true.

---

[92] Billy Bryson, *A Short History Nearly Everything*, (Broadway Books, New York. 2003). 76

Health bureaucracies

Turn over all health records to the individual customer. In a rational world there are no "patients", only customers. And medical doctors who serve them by "performing" rather than "practicing". The practice is done before they become a doctor, after which they become a performer.

Individual doctors and hospitals may and should keep copies. But only by agreement with the customer. I keep my own records. Or rather my wife does that for me. This includes x-rays, scans, lab reports, prescriptions and so on. No one else has a copy. When we go to see a medical practitioner—who is always at the stage where they are performing—my wife has the file nicely arranged for the performer. As a "cash customer" we can do this. I plan to have it available online in the near future.

If a health insurance company feels they can make the transition to a free market economy then they should keep copies of records as appropriate, with the permission of the customer. In the long-term retaining records only for those who chose to remain customers of the insurance company.

You should read my essay on *Medical Doctors Should Perform, Not Practice*. See the reader section of the publisher web site. <http://universal-salvage.org.uk/books/>.

Postal services

In the future even so-called "first class" mail would be handled by private companies. As private companies now handle almost everything except first class. A bureaucratic post office is another of those non-essential, antediluvian operations not needed in a world of individual freedom.

Someone should rename their current mail company the "American Letter Mail Company" or the "Spooner Mail Service". To honor Lysander Spooner, who operated a private mail company in the 1840s until he was forced out of business by the American political régime.[93] Seems his prices were too low and his service too high for the postal monopoly to compete with him. And they did want a monopoly. Simple solution? Create an unnatural law that made it illegal for a private company to deliver letters. And force him to stop under penalty of death. The usual penalty for violating any unnatural law.

If you don't believe this, read my essay *Cat License or Death* at:
<http://www.aristarchus.org/Cat-License-or-Death.php>

The war on drugs

At $40 billion/year a lot of money can be saved by eliminating this. Which would have no real effect on the availability and usage of drugs. Except perhaps reducing the demand in the near term and eliminating it in the long term. True and complete individual freedom is what will eventually eliminate it. At no cost to individuals in society. The quality of life of non-users will be immediately improved by individual freedom. And of drug users a little later.

In one year that is *twice* the amount of money for the *ten year* plan to "put a man on the moon". Yes, there has been much inflation since the 1960s, but then that is another product of political states.

Life under a societal system based on coercion is so bad that some individuals seek relief from all the problems created by politics. It is unfortunate that some seek relief here, as it is an expensive illusion. The biggest expense not being monetary, but on the qualify of life of the individual. What it does to the long-term survivability of the individual. They begin to lose and increasingly lose the ability to control their own lives.

Most people realize that the "war on drugs" is a futile, costly effort that has no effect on the availability and number

---

93 Lysander Spooner (1808-1887). An individualist who, as I, favored natural law as opposed to unnatural law.

of users. The only real "benefit" is the employment provided to the war makers. Further, it has had an adverse effect on the treatment of pain. You can search on the Internet to find information on the failure of "the war on drugs". It is about as successful as Bush's "war on terror". Using terrorist tactics.

Thinking more of the "war on drugs", I suddenly thought that the availability of drugs and their quality continues to go up. And the price come down. And that this is the opposite of the products of the political state? Where the quality goes down and the price goes up? And fritter away $40 billion per year in the bargain?

Antitrust organizations

What better organization to get rid of than one that opposes trust? The state-controlled monopoly that opposes anything they see as competition.

I can see why the political state opposes any large organization: it represents a threat to their regulatory and other immoral powers. The individuals who operate these organizations represent the height of irrationality, immorality, ignorance, impertinence and insolence. The I^5 mentality.

The natural evolution of a truly successful company is to become a monopoly. Representing an organization with the lowest possible prices, best products and service. The best management. And in a free society can not represent a threat to anyone. It is only in societies based on coercion that there can be an apparent need for a "monopolies commission". As it is only in societies based on coercion that there can be corruption.

Centers for Disease Control (CDC)

The *function* of the American CDC and any equivalent organizations elsewhere has a place in a free world. Diseases will probably change and increase in an ever-hotter world. There may be one small advantage to living with perpetual ice storms. Fewer bacteria?

In a sense this service could operate as an "association"

providing services to insurance and medical organizations. Or itself becoming a medical insurance operation? Perhaps open one or more hospitals? These could provide clinical, research, surgical, and teaching and training facilities

The capabilities and knowledge the organization has must not be lost

Final Remarks

Somehow, neither "summary" nor "conclusion" seemed right. Not sure if this is right either.

You think I already said too much? Not enough? In either case the situation is that individuals must *think*. And do what in their own mind is best for individual survival. Which in most cases—when carefully thought out—will also be good for survival of the specie.

As some old bureaucratic saying went: "Remember, the life you save may be your own."

As long as there is no coercion to use services, patronize only those services you need. The others will whither away.

I know: politics has so warped the fabric of society that it does seem extraordinarily difficult to disentangle ourselves from it. As I have said:

What a tangled we devise
when first we subsidize.

Even the best minds, such as Richard Branson, could not seem to make British Rail work after decades of socialist mismanagement. The American regime can not really make Amtrak work. But then maybe this is because they still keep their finger in the pie? I can see many things working very easily very quickly. Then there are the myriad of things that don't even need to be done. Such as marriage and business licenses, customs and immigration, zoning, etc.

Review

Let's review the reason why we have never solved any of our societal problems with a system based on coercion. Based on the concept of politics. Why we can never solve any of our societal problems using this system. Why we must have a societal system based on total individual freedom. And if we do not do this, we will not solve the societal problems resulting from Global Warming. And why we will thus become extinct as a specie. Taking with us all life on this planet except the simple organisms with which we started.

All members of the specie Homo sapiens are unique and different from all others of the specie. This is a law of nature. The laws of nature are immutable and can not be contravened.

A societal system based on coercion seeks to contravene a law of nature. To make a mockery of the concept of thinking life. Of thinking beings taking action based on their thoughts. And being responsible for the results of that action. Learning from their mistakes and becoming more fit as individuals. Thus creating an ever-better life for themselves as individuals. And as a natural byproduct, for all other members of the specie.

When individuals are in charge of their own lives and must live the results of their actions, they have an interest in not creating conflict. As it may not be good for them. But politicians have no such restriction or "regulation" on their life. As if things go wrong it does not effect them. And they can blame someone else. And then be promoted to their next level of incompetence.

There are some "books of inspiration" that can help you:[94]

- *The Discoverers*, by Daniel J. Boorstin.

<universal-salvage.org.uk/other_books.php#boorstin>

- *A Short History of Nearly Everything*, by Bill Bryson.

<universal-salvage.org.uk/general.php#short>

- *The Mainspring of Human Progress*, by Henry Grady Weaver.

---

94 This links will first present you with reviews and other information on the books. And then a link to Amazon to buy the book if you wish.

<universal-salvage.org.uk/other_books.php#weaver>

- *Mysterious Island*, by Jules Verne.

<universal-salvage.org.uk/swift_verne_books.php#swift>

There are probably many others. But I am most knowledgeable on these and they are very appropriate. Books illustrating the value of individual freedom.

In *The Discoverers*, Boorstin tells the story of the *intellectual voyage* of mankind. How we acquired new and useful knowledge to learn about our world. How we used that to create tangible products. And who did it.

Bill Bryson in his *A Short History of Nearly Everything* carries on with this, but covering mostly recent times. The past two or three hundred years. And provides some very, very interesting information about our world.

The Mainspring of Human Progress of Weaver tells a story of individual freedom and what it did and can do for individual Homo sapiens. It discusses the great Saracen (Arab) civilization that preserved the knowledge of the Ancient Greeks. And then in the 1200s the translation of these into Latin, bringing new and useful knowledge to Europe. Knowledge that began to end the Dark Ages and the Inquisition.

Mysterious Island by Verne may seem an odd and whimsical choice from childhood. But it tells the story of a few men marooned on an island about which they knew nothing. Yet they had an unusual knowledge of the laws of nature. And acted as individual thinking beings. To single handedly almost create a "modern, technological world".

To me the most impressive part of *Mysterious Island* is that Jules Verne had to have the knowledge of nature contained in the minds of those stranded on Mysterious Island. Knowledge that almost no one had at the time Verne wrote his book.

Increasingly in the UK and even the USA *individuals* are sensing the reality of the situation and are *individually* taking action to do something about Global Warming. But this is something that must increase dramatically. Where people are saying that "Well, the politicians do not understand and they do nothing. Screw it. I will do something". The politicians are "excess baggage". And if you travel very much you must know that excess baggage is a very expensive thing.

We must think about how to handle the societal effects of Global Warming. Which can only be done through changes in our societal structure. Through solving the problem of *Global Dumbing.* And about how to reduce particulate and greenhouse gas emissions. How to add oxygen to the atmosphere and remove carbon dioxide. Innovations that can do this and even begin to bring temperatures back to about where they were. It may take centuries, even millennia to undo the damage we have done in less than one hundred years. We can do it if we become Homo sapiens sapiens. But we must do that *now.*

The plan, the solution

And yes, the plan and the solution. Well, you see, this is very simple. If you have been thinking as you read it is obvious. To solve the problems of Global Dumbing and then Global Warming there are only two simple things to do.

1) Assume responsibility for your own life.
2) Take those actions which are good for you in *the long term.*
3) Acquire job security for yourself.

By definition, and consistent with the laws of nature this—and only this—will solve our problems. For some people this may take days. For many it will take longer. It depends only on the understanding and agreement of other individuals. If perhaps 2% of the people understand and start to do it then we have a chance of not becoming extinct.

Appendix A

Definitions

Version 3: 2006, October

Many of these definitions were first proposed by Andrew J. Galambos. To whom I owe a tremendous amount of gratitude for their creation and for making them available to me. I have added a number of definitions created by myself. To the extent the definitions are correct this should be to the credit of Galambos. When a definition is proven to not be correct that should be for my account.

Absolute

"That which is independent of arbitrary standards of measurement".

Aristarchus

"The proper name for the planet currently referred to as Earth".

With credit to Andrew J Galambos (1924-1997), the creator of this idea. Aristarchus of Samos (310-230 bce) was one of the first to accurately determine the size of the planet.

Bias

"A preference based on discrimination".

See "Discrimination" and "Prejudice" below.

Bush War

"A preemptive military strike in which the aggressor loses the war and the country attacked descends into civil war. Which is hailed as a great victory by the aggressor".

George Orwell would probably call this a "double." The Ignorant One has engaged in both Doublethink and confirmed Orwell's statement that "politics is the defense of the indefensible." With *Doublethink* meaning the ability to hold two contradictory concepts in the mind at the same time and accept both of them.

Capitalism

"The societal structure that protects all forms of property completely".

Coercion
"Any attempted and intentional interference with the property of another individual".
Communism
"A quasi-economical system under which the political state controls the lives and property of all the people and claims to own all property".[95]
See "Socialism" below.
Competence
"A measure of the number of things tried related to successful outcomes.".
I have proposed the *Shackleton* as the unit of competence.[96] With 10 Shakletons being total, complete competence. Maybe in the same sense as the Carnot Engine. Representing perhaps an unattainable objective?
Confidence
"Belief that something will happen based on rational reasons and valid thought processes".
Ant. *Faith*
Copernicus
"The proper name for our local star, which we currently refer to as 'Sun'".
As proposed by Andrew J. Galambos (1924-1997), the creator of the thought. In honor of Nicolas Copernicus. (1473-1542), the creator of heliocentric hypothesis.

---

95 Ownership is a moral concept and can not be achieved through coercion or fraud. By immoral means. Thus those who claim to own property but merely control it—such as the political state—do not own it. As they acquired their control through immoral means.

96 In honor of Ernest Shackleton, the great Antarctic explorer. A man of supreme competence.

Crime

"A successful act of coercion".

Crime Against Humanity

"The societal condition that exists when one individual is allowed to interfere in the property of another or the act of doing so".

The problem is that once this interference in and control of the lives of others is a agreed to be acceptable, there is no limit and end to it.

Critic

"A person who complains about the way something is done or the result but who can propose no better way of doing it and/or improving the result".

A critic is one of the more despicable classes of people.

See "Reviewer" below.

Democracy

"A political system wherein the ruler and ruling party faction are replaced by majority vote".

Diplomacy

"Organized lying".

Discrimination

"The act of choosing or selecting".

Thus, any person who does not discriminate is not thinking.

See "Bias" above.

Dictator

"Anyone who controls the life and property of another thinking being".

Thus, all politicians are dictators.

Dictatorship
"A political system wherein the ruler is replaced by force".
Equality (in a societal sense)
"Born and existing under the laws of nature, and having unrestricted access to nature".
Fair
"Any act that does not involve coercion or fraud".
Faith
"Belief that something will happen based on irrational reasons and invalid thought processes".
Freedom
"The societal condition that exists when every individual has 100% control over their own property".
Good (Absolute)
"Anything is an absolute good that is a subjective good to at least one person, but which does not involve the exercise of coercion".
Good (Subjective)
"Any phenomena, process, or action that an individual prefers."
Government
"A mechanism that provides services that protect property."
History
"A presentation of past events."
See "Positive History" below.
Hypothesis
"A suggested explanation for the cause of an event or phenomenon."
It may or may not be right on an absolute basis, and thus become a theory. Hypotheses become theories when validated through the epistemological concepts and steps of what is commonly known as "The Scientific Method".
As outlined by Galambos, adding the concept of "rightness" to The Scientific Method allows it to be used in validating knowledge concerning volition.
Humanist
"Someone who believes in and lives by the ethical doctrine of humanism."
See "Humanism" below.

Humanism
"The ethical doctrine that each individual is responsible for the moral conduct of their own life."
Human Rights
"The individual's right of freedom of access to the universe and the sanctity of property."
Idea
"An articulated thought".
Illegal
"That which a political state decrees or states is not legal".
The concepts of legality and illegal are intellectual fiction. In an advanced, rational, and durable society the words would only be known to some historians. People in general would not believe the concept if you tried to explain it to them.
Impossible
"Any contemplated or attempted action that would result in the violation of a law of nature".
See "Possible" below.
Importance
"A measure of the total amount of property created".
Injustice
"When there is a crime for which there is no recourse".
Intellectual (An)
"A thinking being who trends toward rational thought".
This of course eliminates most people commonly thought to be "intellectuals". And elevates many of those not thought to be an "intellectual" to that position.
Intellectual Honesty
"The ability to accept a theory which—by definition—is right on an absolute basis, when it conflicts with currently held beliefs on the subject".
Justice
"The condition that exists when all injustices are eliminated".
Kepler
"The proper name for the planet we currently refer to a Mars".
As proposed by Andrew J. Galambos (1924-1997), who is the creator of the thought. In honor of Johannes Kepler(1571-1630)

Leader
"Any individual who morally creates new and useful knowledge or products that have the potential for increasing the quality of life for other individuals."
Legal
"That which a political state permits a person to do."
Level Playing Field
"The societal condition that exists when anyone can do anything they wish that is moral."
The concept is the cherished objective of altruists worldwide. But perhaps not this definition? Which is a proper, right definition. See "Right" below.
Libertarian
"A person who believes in total freedom for each and every individual".
Liberty
"The societal condition that exists when all individuals have total control over their own property."
Monarchy
"A political system wherein the ruler is usually replaced by a family member upon the death of the current ruler. Either natural or otherwise."
Money (Political)
"A medium of exchange created by or the use of which is authorized and required by a political state".
Money (Rational and Moral)
"A medium of exchange used in a free market".
Moral
"Any action that does not involve intentional interference in the property of another individual".
Nation
"A political entity".
National
"Of or pertaining to a nation".
From the Latin *nation*, basically meaning an ethnic group or "tribe". A tribe being what originally separated a group of people and made them different from each other.

Nationalist

"One who believes in and practices nationalism".

Nationalism

"The failed unethical doctrine that the only viable societal system is one based on coercion. With a ruler and ruling party faction to control the lives and property of all".

Natural

"Pertaining to the laws of nature".

Non-Volitional Sciences

*"The study of the area of knowledge **not** related to the societal structures of a volitional specie".*

Oberth

"The proper name for our local mono".

With credit to Andrew J. Galambos (1924-1997), the creator of the thought. In honor of Hermann Oberth (1894-1989), who in the 1920s designed a liquid-fueled rocket that could take us there, and proposed a design for a Lunar Excursion Module. Oberth is one of the great overlooked geniuses of the 20th century.

Patriot

"Anyone who adds to the quality of individual life without interfering in the property of another thinking being".

See "Traitor" below.

Play

"The act of self-enjoyment performed after the act of production".

See "Production" and "Re-Creation" below.

Policy Maker

"Someone who conspires to create unnatural laws that will produce interference in the property of another thinking being".

Political System

"A societal system based on coercion".

Positive History

"The version of a historical event that tells the true cause and result of the event".

Possible

"Any proposed act or action that does not violate a law of nature".

Please note that this has nothing to do with the probability or plausibility of something happening. Or of the knowledge related to how to do something.

Thus, electricity and computers were not impossible during the time of the Roman Empire, It is just that no one had yet discovered and elucidated the laws of nature related to the concepts.

Prejudice

"A previously held bias".

See "Bias" and "Discrimination" above.

Production

"The act of creating tangible or intangible property".

Profit

"Any moral increase in happiness".

Isn't that definition created by Andrew J. Galambos beautiful? It brings back to respectability a word that has fallen into disrepute. A nice, simple statement of "Thank you" for an action can be profit. As can a smile or a sunset.

Property (General)

"Individual man's life and all non-procreative results thereof".

This carries with it the concept that children are not property. Rather, a parent has a contractual relationship with a child. To nurture, educate and support them to the point where they can independently perform the responsibilities and obligations of individual thinking life.

Primordial Property

"Individual life itself".

The physiological body of an individual thinking being. What I like to think of as "the environmental support module for the brain". Which of course includes the brain…

Primary Property

"The thoughts of an individual that become ideas."

Beliefs, opinions, theories, the concept behind inventions or products, etc.

Secondary Property

"The tangible result of moral activities of individual life."

Those tangible assets created or acquired through moral action.

Re-Creation

"The act of creating a product".

Tangible or intangible. This replaces the current word and dysfunctional concept of "recreation". Which should now be refereed to as "play". See above.

Reviewer

"A person who creates and presents a non-critical review of something".

See "Critic" above.

Right

"That which is based on true premises, valid and logical thought processes, in which the resulting action is moral".

Science

"The totality of all organized, right knowledge".

Slavery

"Seizure or control of the property of an individual without their consent".

Or,

"The societal condition that exists when someone controls the life and property of others".

Socialism

"A quasi-economic system under which the political state controls the lives and property of all the people and claims to own most of the property".

See "Communism" above.

Sovereign Entity

"And individual who has total control over their own property".

Stealing

"The seizure of the property of an individual without their consent".

Terror

"A feeling of intense fear".

Terrorism

"Intentional interference in the property of another individual".

After all, might this not induce terror in the victim? And how much interference is required for an attack on property to constitute "terrorism"? One-third of a person's property? Or 100%? Or the property of more than two people? Or more than seventeen people? Thousands? Millions?

How about simplifying it as the definition states?

Terrorist

"Any individual who creates a feeling of terror in another individual".

Theory

"The explanation for the cause of a phenomenon or event that is right on an absolute basis".

Traitor

"Anyone who follows the dictates of a political state or its ruler".

Treason

"Any act that interferes in the property of another thinking being".

True

"That which is observable".

With the senses, such as vision, hearing, smell, taste, etc. And in order to be "true" it must be observable by any other person in the same place at the same time under the same conditions.

Utopia (normal)

"A world in which each individual may do anything that does not intentionally degrade the life of another individual".

Thus, all their hopes, dreams and desires for their life may be realized consistent with their ability to do so.

Utopia (political)
"A mystic world in which the individual need do nothing and all their desires for life will be realized".
This is a faulty, failed, impossible-to-implement concept. It depends on the violation of one or more laws of nature.
Valid
"Having a conclusion derived from true premises".
Volition
"The characteristic of being able to reason, to make choices and value judgments, and to take action based on that process".
Volitional Science
"The study of the area of knowledge related to the societal structures of a volitional specie".
Commonly and imprecisely called "social sciences" which, as practiced on this planet at this time is neither "social" nor "science".
Voting (moral)
"The act of selecting a product or service which you can return or dispose of when and if you wish, and which will not have an adverse effect on the lives of others".
Voting (political and immoral)
"The act of selecting a product or service which is virtually impossible to return or dispose of when and if you wish, and which will have an adverse effect on the lives of others".
War
"An organized, legal conspiracy to engage in terrorism".
It is organized, isn't it? By definition it is legal. While dictionaries use modifiers such as "secret" and "unlawful," the basic root meaning is to "breath together." To be together and to agree together to do something.
But does any of this change the principle and the morality? Once it is agreed that under certain circumstance it is OK to kill other people and destroy their property you have opened a "Pandora's box" of immorality and obfuscation. Which can only increase.

War Lord
"Anyone who works for a political régime and controls those known as 'soldiers' or 'fighters' whose mission is to destroy people and property".

You may view the current, possibly revised and augmented definitions at:
<http://universal-salvage.org.uk/books/>.

Appendix B

The Quotations

Version 1: 2006, April

I believe that simply by reading these and sitting back and thinking about them a person should be able to understand much more about our societal structure. About history. And begin to see the true cause of our societal problems. Maybe even the cure.

I have many other interesting, useful, though-provoking quotations in my database. And I am sure there many others I do not know about and do not have.

The following are not arranged in any order…

Victor Hugo (1802-1885)

> "*There is nothing so powerful as an idea whose time has come*".

John Locke (1632-1704)

> "*New ideas are always suspected, and usually opposed without any other reason but because they are not already common*".
>
> An Essay Concerning Human Understanding (1690, dedicatory epistle)

Isaac Newton (1642-1725)

> "*If I have seen further than others,*
> *it is because I have stood on the shoulders of giants*".
>
> Letter to Robert Hooke - 1676/02/05

"I do not know what I may appear to the world,
but to myself I seem to have been only like a boy
playing on the sea-shore and diverting myself
in now and then finding a smoother pebble or a prettier shell than ordinary, whilst the great ocean of truth lay all undiscovered before me".

from Brewster, *Memoirs of Newton*, [1885] vol II, ch. 27

As Galambos pointed out, Newton's must important creation was not his laws of motion or optics. But rather his statement that "… the universe is comprehensible". And directing us to the epistemological pathway to comprehend the universe. That of the Scientific Method. Which also allows us to validate existing knowledge. To determine what is right in the non-volitional world.

Then Galambos created a farther account. Saying that simply by checking the morality of any action we can apply the Scientific Method to the solution of societal problems. A "simple" but very elegant and important concept.

Albert Szent-Gyürgyi von Nagryapolt (1893-1985)

"Discovery consists in seeing what everyone else has seen,
and thinking what no one else has thought".

Source not known

Thomas Paine (1737-1809

"A long habit of not thinking a thing wrong,
gives it a superficial appearance of being right,
and raises at first a formidable outcry in defense of custom.
But the tumult soon subsides.
Time makes more converts than reason".

Thomas Paine, Introduction to *Common Sense*, Penguin Books, 1983, p. 63

"O ye that love mankind! Ye that dare oppose, not only the tyranny but the tyrant, stand forth! Every spot of the old world is over-run with oppression. Freedom hath been hunted round the globe. Asia, and Africa, have long expelled her. Europe regards her like a stranger, and England hath given her warning to depart. O! receive the fugitive, and prepare in time an asylum for mankind".

Common Sense [1776], by Thomas Paine

* * * * *

"In England a king hath little more to do than to make war and give away places; which in plain terms, is to impoverish the nation and set it together by the ears. A pretty business indeed for a man to be allowed eight hundred thousand sterling for, and worshipped into the bargain! Of more worth is one honest man to society and in the sight of God, than all the crowned ruffians that ever lived".

Common Sense, 1776

* * * * *

"The heathens paid divine honours to their deceased kings, and the Christian world hath improved on the plan by doing the same to their living ones. How impious is the title of Sacred Majesty applied to a worm, who in the midst of his splendour is crumbling into dust".

Common Sense, 1776

* * * * *

"Men who look upon themselves born to reign, and others to obey, soon grow insolent; selected from the rest of mankind their minds are early poisoned by importance; and the world they act in differs so materially from the world at large, that they have but little opportunity of knowing its true interests, and when they succeed to the government are frequently the most ignorant and unfit of any throughout the dominions".

Common Sense, 1776

* * * * *

"It is an affront to truth to treat falsehood with complacency".

Source not known.

"Taxes are not raised to carry on wars, but wars are raised to carry on taxes."

Source not known

*****

"These are the times that try men's souls. The summer soldier and the sunshine patriot will, in this crisis, shrink from the service of his country; but he that stands it now, deserves the love and thanks of man and woman. Tyranny, like hell, is not easily conquered; yet we have this consolation with us, that the harder the conflict the more glorious the triumph. What we obtain too cheap, we esteem too lightly; 'tis dearness only that gives everything its value. Heaven knows how to put a proper price on its goods; and it would be strange indeed if so celestial an article as Freedom should not be highly rated".

The Crisis, by Thomas Paine. December 23, 1776.

Jonathan Swift (1667-1745)

"When a true genius appears in the world, you may know him by this sign, that the dunces are all in a confederacy against him".

Hermann von Helmholtz (1821-1894

"The originator of a new concept...finds, as a rule, that it is much more difficult to find out why other people do not understand him, than it was to discover the new truth".

Source not known

George Bernard Shaw (1856-1950)

"The reasonable man adapts himself to the world, but the unreasonable man tries to adapt the world to him—therefore, all progress depends on the unreasonable man".

Source not known

J. E. E. D. Acton (1834-1902)

"Power tends to corrupt and absolute power corrupts absolutely".

Letter to Bishop Mandell Creighton, 1887, April 3

William of Occam [Ockham] (1300-1349)

"*Entia non sunt multiplicanda præter necessitatem*".
[No more things should be presumed to exist than are absolutely necessary, or, Entities should not be multiplied unnecessarily]

Quodlibeta Septem, [c. 1320]

Francois Marie Arouet [Voltaire] (1694-1778)

"It is forbidden to kill; therefore all murderers are punished unless they kill in large numbers and to the sound of trumpets".

Philosophical Dictionary

"Si Dieu n'esistait pas, il faudrait l'inventer".
[If God did not exist, it would be necessary to invent him.]

Source not known

Publius Cornelius Scipio (236-183 bce)

"*No man should rule other men, and certainly not for a lifetime*".

Said when, as a result of his victory over Hannibal at Zama in 202 bce during the Second Punic War, he was offered the rulership for life of the Roman Empire. The empire he had created. He turned it down, accepting only for the normal one year period.

Christian Morgenstern (1871-1914)

"*Das Offensichtliche ist das, was man nicht sieht, bis jeman es erklärt*".
[The obvious is that which is not seen until someone first explains it clearly]

Source not known

Richard Greenspan (dates unknown, current era)

"*People who have learned something as the truth don't particularly like to hear that they've based their careers on things that were either incomplete, or not completely correct*".

Gary Taubes, "An Electrifying Possibility," *Discover*, April 1986, p. 37

Anatole France (1844-1924)

"Ignorance is the necessary condition of life itself. If we knew everything, we could not endure existence for a single hour".

Jules Henri Poincaré (1854-1912)

*"Science is built up with facts, as a house is with stones.
But a collection of facts is no more a science than a heap of stones is a house".*

Benjamin Disraeli (1804-1881)

"To be conscious that you are ignorant is a great step to knowledge".

Herbert Spencer (1820-1903)

"Science is organized knowledge".

William W Morgan (Contemporary…)

"The limits of growth are found in the minds of small men".

*****

"A prayer a day keeps rationality away".

*****

"In true capitalism there are no guarantees of success. There is only the guarantee of the full, unhindered right to try for it, and be compensated according to the results".

*****

*"What a tangled web we devise,
when first we subsidize".*

"Freedom and democracy are mutually exclusive concepts. In a political society based on the concept of democracy, true individual freedom can not exist. In a free society, the concepts of a political society based on democracy would be unthinkable. There would be no market demand for the concept".

* * * * *

"The Grail for which the Knights of the Roundtable searched--and for which all volitional beings search--is not just truth as they thought, But truth combined with validity and morality, which then become rightness".

This thought occurred to me while watching "Excalibur," John Boorman's interpretation of Malory's Arthurian legend.

* * * * *

"Generally speaking, the importance of a public speaker bears an inverse relationship to the number of microphones into which he or she speaks".

* * * * *

"The longer and more extensive is the interference in the free market process, the more difficult and traumatic is the return to freedom, sanity, and morality".

* * * * *

There is a direct connection between morality in a volitional society and individual physiological immortality".

"There is an inverse relationship between the importance of a new idea, and the ease and time in which it is understood, accepted, and implemented".

"The supreme act of fraud and witchcraft for politicians and the political mechanism--and especially for proponents and apologists for the concept of democracy--is in controlling the lives and property of their victims while at the same time having them believe they have freedom and are free".

* * * * *

"Belief in the concept of democracy as the way to freedom requires the acceptance of the idea that freedom can only be obtained through the mechanism of a minority of people controlling the lives and property of others".

This conforms to the Marx, Engels and Lenin concept, yes? When they said the political state would "wither away" and there would be the "rulership of the proletariat". Which I think is the same as "majority rule"?

* * * * *

"The chaos and disorder that will occur on a society is inversely proportional to the amount of individual freedom in that society".

* * * * *

"There is a direct relationship between the attempt to eliminate uniqueness and the attempt to impose equality in volitional societies, and the level of frustration and violence that will ensue".

* * * * *

"In a Monarchy the chant is: The King Is Dead. Long Live the King;
In Democracy the chant is: The Incumbent is Out: Long Live the Incumbent".

Under both concepts the individual is out but the system continues. How many Kings have there been in history? How man Political Presidents? How many societal problems have we solved under the system of regulating individual behavior through coercion?

“As the number of sovereign entities in an area approaches the total population something interesting happens: It is called freedom”.

* * * * *

There is only one job security: ”The ability of an individual to perform a task or tasks currently in demand in the marketplace at an acceptable market price”.

* * * * *

‘Production compensation must be based solely on the market value of error-free production”.

Henry Fielding (1707-1754)

"*He in a few minutes ravished this fair creature, or at least would have ravished her, if she had not, by a timely compliance, prevented him*".

> [This reminds one of the poor taxpayer and "voluntary" compliance. Except with taxes you are presented with a false alternative. You are ravished in both cases, even when you do pay "voluntarily"]

If you are interested you can access my quotations database at:

<http://universal-salvage.org.ui/quotations>

I can’t guarantee it will always work, but at the least this address should provide some information on it…

Appendix C

Bibliography

You may view this bibliography on-line as an HTML document and thus be able to simply click on the book links. Also, if there are later revisions to this bibliography they will be available there. The link: <http://universal-salvage.org.uk/books/>

This will take you to the publisher's "Reader Index," from which you may access book revisions, bibliographies and other such things.

The Mainspring of Human Progress
By Henry Grady Weaver
The Foundation for Economic Education,
Irvington-on-Hudson, NY 105333
ISBN: 1-57246-064-4
<http://universal-salvage.org.uk/other_books.php#main>

Brave New World
By Aldous Huxley
Perennial Classics, HarperPerennial
First Perennial Library Edition, 1966
ISBN: 0-06-09287-1
<http://universal-salvage.org.uk/huxley_orwell_books.php#brave>

Nineteen Eighty-Four
By George Orwell
A Signet Classic, Penguin Books Ltd, England
ISBN: 0-451-52493-4
<http://universal-salvage.org.uk/huxley_orwell_books.php#1984>

The Discoverers
By Daniel J. Boorstin
Vintage Books, Random House, NY
ISBN: 0-394-72626-1
<http://universal-salvage.org.uk/other_books.php#boorstin>

Natural Right and History
By Leo Straus
University of Chicago Press
ISBN: 0226776948
<http://universal-salvage.org.uk/general.php#leo>

Evolutionary Origins of Morality
Edited by Leonard D. Katz
Imprint Academic (March 1, 2000)
ISBN: 090784507X
<http://universal-salvage.org.uk/general.php#morality>

The Discovery of Freedom
By Rose Wilder Lane
50th Anniversary Edition (1993)
Fox & Wilkes, San Francisco, California
ISBN: 0-930073-21-5
<http://universal-salvage.org.uk/other_books.php#free>

Economics in One Lesson
By Henry Hazlitt
Three Rivers Press, NY
ISBN: 0-517-54823-2
<http://universal-salvage.org.uk/other_books.php#econ>

A Greater Than Napoleon
By B. H. Liddell Hart
Da Capo Press, New York
ISBN: 0-306-80583-9
<http://universal-salvage.org.uk/scipio_by_about.php#hart>

Cowardly New World
By William W Morgan
Universal Salvage Corporation, Escondido, California
ISBN: 1-4196-2610-8
<http://universal-salvage.org.uk/book_list.php>

Colours of Infinity: The Beauty and Power of Fractals
By Nigel Lesmoir-Gordon
Clear Books
ISBN: 1904555055
<http://universal-salvage.org.uk/general.php#colours>

The Journey of Man: A Genetic Odyssey
By Spencer Wells
Random House
ISBN: 0812971469
<http://universal-salvage.org.uk/general.php#journey>

The Image: A Guide to Pseudo-Events in America
By Daniel J Boorstin
Vintage Press
ISBN: 0679741801

The Seekers
Daniel J. Boorstin
Vintage Books, A division of Random House (1999)
ISBN 0-375-70475-2
<http://universal-salvage.org.uk/general.php#seekers>

A Short History of Almost Everything
By Bill Bryson
Broadway Books, NY (2003)
ISBN: 0767923227
<http://universal-salvage.org.uk/general.php#bryson>

The Internal Constitution of the Stars
By Arthur Stanley Eddington
Cambridge University Press
ISBN: 0521337089
<http://universal-salvage.org.uk/eddington_by_about.php#internal>

Stellar Structure
By Arthur Stanley Eddington
<http://universal-salvage.org.uk/eddington_by_about.php#stellar>

Sir Isaac Newton: His Life and Work (Paperback)
by Edward N. Andrade
Publisher: Doubleday (January, 2000)
ISBN: 038502326X
<http://universal-salvage.org.uk/newton_by_about.php#sir>

The Life of Sir Isaac Newton
By David Brewster
<http://universal-salvage.org.uk/newton_by_about.php#life>

A God Against the Gods
by Allen Drury
ISBN: 0385001991
<http://universal-salvage.org.uk/other_books.php#allen>

An "Einstein" Thought Experiment: Rationality & Morality Waves in Nature
By William W Morgan
ISBN: 0-9778492-0-1
<http://universal-salvage.org.uk/book_list.php>

Sic Itur Ad Astra
By Andrew J Galambos
ISBN: 0880780045
<http://universal-salvage.org.uk/galambos_pubs.php#isc>

You may view this bibliography on-line as an HTML document and thus be able to simply click on the book links. Also, if there are later revisions to this bibliography they will be available there. The link:
<http://universal-salvage.org.uk/books/>

This will take you to the publisher's "Reader Index," from which you may access book revisions, bibliographies and other such things.

Appendix D

Achievements

This list of Seekers is nor intended to be comprehensive. The fact that a name is not on the list does not mean the individual is not important in the effort to understand Global Warming and to overcome the problems that will result from it.

Please note there is only one politician in the list. And he is there not because he is a politician, rather in spite of that fact. He has done and is doing something positive related to Global Warming. He in fact does understand the problem and is pro-active in that, rather then being merely reactive.

Other than that, politicians and politics are part of the problem and not part of the solution.

I intend to have this list available on a web site with active links to information about some of the people. In the future it should have more links than are listed here. The address is: <universal-salvage.org.uk/heroes.htm>

Again, please remember that if a person is not on the list it does *not* mean they are not important as part of handling the problems of Global Warming and Global Dimming. Unless of course they are politicians. No politicians will be listed on this list. I am promoting Mr. Albert Gore to Homo sapiens.

The following are in random order...

Dr Peter Cox
Hadley Centre
Climate Prediction and Research
Peter Cox and his team were the first in the world to include vegetation and carbon cycle feedbacks in a climate prediction model. Their results have shown that these feedbacks could significantly accelerate Global Warming.

Graham Farquhar
Australia
Visit the following URL for additional information:
< http://www.rsbs.anu.edu.au/profiles/graham_farquhar/>

Dr Hayley Fowler
Senior Research Associate
School of Civil Engineering and Geosciences
Newcastle University
<http://www.staff.ncl.ac.uk/h.j.fowler/>

Albert Gore
Producer, *An Inconvenient Truth*
Author, *An Inconvenient Truth*

Dr Beate Liepert, Doherty Research Scientist
Columbia University
Lamont-Doherty Earth Observatory
Palisades, New York
<http://www.ldeo.columbia.edu/~liepert/>

Prof Veerabhadran Ramanathan
University of California at San Diego
Victor Alderson Professor of Applied Ocean Sciences
Professor of Climate and Atmospheric Sciences
<http://www-ramanathan.ucsd.edu/>

Dr. Michael L. Roderick
Australia National University
Research School of Biological Sciences
Ecosystem Dynamics Group
Canberra, Australia
<http://www.rsbs.anu.edu.au/Profiles/Michael_Roderick/>

Dr. Leon Rotstayn
CSIRO Marine and
Atmospheric Research
Castray Esplanade
Hobart, Tasmania
<http://www.cmar.csiro.au/>

Gerald Stanhill
Agricultural Research Organisation
Department of Environmental Physics and Irrigation
Institute of Soils, Water and Environmental Sciences
Agricultural Research Organization,
The Volcani Center,
<http://www.agri.gov.il/People/GeraldStanhill.html>

Benjamin Santer
Climate Researcher
Lawrence Livermore National Laboratory
<http://en.wikipedia.org/wiki/Benjamin_D._Santer>
< http://www.llnl.gov/pao/news/news_releases/2005/NR-05-10-01.html>

Chris Stringer, Merit Researcher Human Origins
Paleontology Department
Natural History Museum, London
<http://www.nhm.ac.uk/research-curation/staff-directory/palaeontology/cv-5508.html>

David Travis
University if Wisconsin Whitewater
College of Letters & Sciences
<http://www.uww.edu/cls/success/faculty/travis_david.html>
University of Wisconsin Bio
<http://facstaff.uww.edu/travisd/>

Katey Walker
University of Alaska
Policy Education Awards:
<http://www.farmfoundation.org/nppechildreth.htm#17>
Methane bubbles climate trouble:
<http://news.bbc.co.uk/1/hi/sci/tech/5321046.stm>

Dr. Eric Wolff, Glacier Chemist
British Antarctic Survey
<http://www.antarctica.ac.uk/staff-profiles/template.php?user=ewwo#biography>

Prof Chris Thomas
University of York
Department of Biology (Area 18)
University of York
YO10 5YW
Tel: 44-(0)1904 32 8646
Fax: 44-(0)1904 32 8505
<http://www.york.ac.uk/depts/biol/staff/edt.hem>
<http://universal-salveage.org.uk/ChrisThomas.htm>
Climate change, habitat fragmentation, conservation biology, metapopulation dynamics, sustainability, butterfly ecology.

Christopher Field, Director
Global Ecology
Carnegie Institute
Washington, DC

Prof. Ted Schuur
Professor, Ecosystem Ecology
University of Florida
<http://ecology.botany.ufl.edu/ecosystemdynamics/schuur/Schuur.html>

Index

9

9-11
- a product of democracy · 93
- time to build World Trade Towers · 27

A

Abraham · 51, 57, 77
Acton, J. E. E. D. · 198
Akhenaten · 58
Alexander of Macedon · 29, 35, 75
Alfonso el Sabio · 35
Americans
- and fighters · 4
- and soldiers · 4

Archimedes · 29, 74, 79, 98
Aristarchus of Samos · 27, 29, 79
axis powers · 75

B

Beatles, the · 74
Beckett, Samuel · 148
black market · 44
Boorstin, Daniel J
- *Hidden History* · 83
- *The Discoveres* · 180

borders, political & natural · 25
Brahe, Tycho · 29, 38
Bronowski, Jacob
- *The Ascent of Man* · 23
- the man · 23

Bruno Galaxy · 64
Bruno, Giordano · 38, 39, 54
Bryson, Bill
- *A Short History of Nearly Everything* · 180

Bush, George W
- a crime to make him look bad · 39
- and collateral damage · 6
- and *Dear Leader* · 34
- and his weapons of mass destruction · 34
- and the Führerbunker · 122
- and the murder of 100,000 people · 6
- and the Wolf's Lair · 122
- and torture · 6
- as *Dour Leader* · 34, 56, 64, 75, 84, 138
- Bush Thought · 89
- rule by decree · 34, 70
- use of weapons of mass destruction · 37

C

Carlson, Chester · xi
Carson, Rachel · 3
Châtelet, Émilie du · 28
Chomsky, Naom · 45
Christ, Jesus · 59, 79
- his biography · 79

civilizations
- dawn of · 28
- Egypt, First Dynasty · 49
- Greeks · 51
- Romans · 52
- Saracens · 53
- Sumerians · 50

clearing house · 159
climate change
- the reality · 1
- very rapid · 21

climatologists · iii
- listing of · 211

Columbus, Christopher · 53
Copernicus · 2
- our star · 38

Council of Nicaea · 32, 83
Cox, Peter · 211

credit system
    arbitration · 117
    as a societal regulator · 86, 113
    credit restoration · 117
    general · 113
    restitution · 116
credit systems · 114
crusades · 61

D

Darwin, Charles · 66, 89
definitions
    and effective communications · 8
    democracy · 66
    dictatorship · 66
    importance of · 8, 9, 33
    monarchy · 66
    political system · 66
discovery
    concept of · 38, 97
discrimination · 25
disorder in society · vi
Disraeli, Benjamin · 200
Drake, Edwin L · 2

E

Einstein, Albert · 74
emission control
    urgency · 146
emission cut-back
    timing · v
essential services
    Centers for Disease Control · 160, 178
    education · 175
    money · 160
    social security · 174
extinction
    point of no return · 2
    the cause · 12

F

Fielding, Henry · 203
Fonteyn, Margot · 74
fossil fuel
    burning of · 3, 6
Fowler, Hayley · 212
France, Anatole · 200
Franklin, Benjamin · 85
Fraquhar, Graham · 212

G

Galambos' Laws · 86, 100
Galambos, Andrew J · iii, 31, 64, 73, 94
    and minimizing entropy · 98
    and the achievement · 95
    and the Carnot Engine of Volition · 98
    and the flow-stream of history · 96
    and the Laws of Volition · 100
Galilei, Galileo · 29, 38
Gang of Four · 75
Gideon · 77
glaciers
    lubrication · 133
    tipping event · 133
global dimming · 4
global dumbing · 2, 149, 170
    SOLUTION · 182
global warming
    and full employment · 131
    and mainstream scientists · 1
    another hidden danger · 126
    changing rainfall patterns · 138
    coastal areas · 125
    desertification · 122
    flooding · 138
    island disappearance · 129
    low countries · 130
    methane hydrates · 123
    real or imagined? · 1
    soaking · 147
    SOLUTION · 182

spontaneous forest fires · 122
storms · 134
tipping event · 122, 132
tipping points · 123
Gore, Al
An Inconvenient Truth · 8
Gore, Albert · 212
Greenspan, Richard · 199
Gutenberg
moveable type printing · 27

H

Hannibal · 84
Helmholtz, Hermann von · 198
Henry the Navigator · 53
Hero of Alexandria · 29, 79
Hewlett, William · x
Hippocrates · 29, 79
Hobbes, Thomas · 88
Hugo, Victor · 195
human evolution
crop cultivation · 27
crop irrigation · 27
metals technology · 27
the wheel · 27
writing · 27
human life
value of · 74
human potential · 156
Huxley, Aldous · 53, 120

I

immigration
as an unneeded function · 140, 141, 143, 159, 172
travel difficulties · vi
individual freedom
and the American colonies · 156
and the USSR · 157
steps to acquire · 158
innovation · 105, 167
Inquisition
first · 34
second · 34
insect life · 139
intelligence
artificial · 113
of nature · 113
Israelis
and mass murder · 4
and Palestinian concentration camp · 4
and world's largest concentration camp · 37

J

jihad · 34, 61, 82
job security · 102, 103
and poverty · 146
and the problems of the Sahel · 152
the urgency · 147
Johnson, Lyndon B · 74

K

Kennedy, John F · 74
Kepler, Johannes · 29, 38

L

Lea, Homer · 17
leaders
definition · 8
false leaders · 8
true leaders · 8
Liddell Hart, B. H.
Scipio Africanus
A Greater Than Napoleon · 31
Liepert, Beate · 212
living
improving our houses · 166
production villages · 167
Locke, John · 195

M

Manilow, Barry · 74
manmade disasters
Amtrak · 4
British Rail · 4
FEMA · 18
Global Dumbing · 12
global warming · 12
healthcare system · 4, 20
London Underground · 4
pension system · 4, 20
school system · 20
school systems · 4
methane hydrates · 2, 3, 123, 139, 149
methane lakes · 149
monotheism · 59
Morgan, William W · 200
Morgenstern, Christian · 199
Muhammad (the prophet) · 59, 61
Muslim Jerusalem
city of religious tolerance · 35
Muslims and religious tolerance · 35

N

Nagryapolt, Albert Szent-Gyürgyi von · 38, 196
Napoleon · 42
natural disasters
asteroids & comets · 11
Carlisle, England · 13, 151
Katrina · 13, 151
magnetic field problems · 11
magnetic fieldproblems · 16
storms · 12
supervolcanoes · 11, 14
nature
balance in nature · 90
beauty of · 113
equality in nature · 90
news
really bad · vii
Newton, Isaac · 74, 195
and the Industrial Revolution · 28, 40
his most important achievement · 40
Principia mathematica · 28
standing on the shoulders of giants · 95
universe is comprehensible · 99
Newton, Isaac · 29
non-essential services
antitrust (monopolies) · 178
Customs · 172
immigration · 172
war (defense) department · 174
war on drugs · 177
zoning · 175

O

Occan, William of · 199
offshore oil platform patented · 3
oil well
first US well · 2
Orwell, George · 45, 57

P

Packard, David · x
Paine, Thomas · 55, 85, 120, 196
Common Sense · 86
Declaration of Independence · 86
The Age of Reason · 86
The Crisis · 86
Pericles · 78
permafrost · 149
personal security
your pension account · 109
your personal pension · 106
Plutarch · 52
Poincaré, Jules Henri · 41, 200
political agent · 65
political licenses
and divorce · 161
and marriage · 154, 161

and travel · 154
driving · 162
general · 162
political overlords
and innovatiion · vi
political police · 4
political regulation
and fire departments · 160
and the clearing house · 159
and the credit mechanism · 112
need for · 68
the air waves · 166
politicians
as followers · 70
politics
and broken news · 7
and doublethink · 6
and false alternatives · 44
and need · 69
and sneakers · 6
and the best wines · 7
and trigger events · 34
as a problem creating system · 7
ever increasing laws · 70
failure of · 4
government without end · 9, 46, 67, 71
loopholes · 43, 48, 93
regulations the Titanic disaster · 43
results of the system · 13
the products · 74
pollution
and our nose · 42
polytheism · 59
prejudice · 25
production compensation · 105
Ptolemy · 29

R

Ramanathan, Veerabhadran · 212
Regulations
moral societal regulator · 101
Renaissance
cause of · 82
Roderick, Michael · 212
Roosevelt, Franklin · 55, 74
Rotstayn, Leon · 213
Rowland, Thomas F · 3

S

Saladin · 63
Samuel · 77
Santer, Benjamin · 213
Saracen civilization · 35
Saracens · 53, 63, 80
science
and the Saracens · 53, 81
Scipio, Publius Cornelius · 30, 51, 84, 199
sea levels
and the fabric of society · v
Sharon, Ariel · 65
Shaw, George Bernard · 198
Shinawatra, Thaksin · x
Silent Spring, by Rachel Carson · 3
Sinatra, Frank · 74
soaking · 147, 168
soaking index · 159
socialism
and the Romans · 52
in the USA · 55, 111
societal chaos
cause of · 42, 56
societal problems · vii
false solutions · 9
societal system
and solving problem of global warming · vii, 1
and the laws of nature · vii
and thinking beings · vii
Speers, Britney · 74
Spencer, Herbert · 200
Spooner, Lysander · 176
Stanhill, Gerald · 150, 213
Strabo · 52
Streisand, Barbra · 74
Stringer, Chris · 213
superpower · 64
supervolcano

Lake Toba · 14
our near extinction · 14
Yellowstone National Park · 14
Swift, Jonathan · 83, 198

T

Tacitus · 52
temperature changes
effects · 19
temperatures
not seen for 4 billon years · 2
safe level · 3
Thales of Miletus · 51
thinking beings
the natural state · 25
Thomas, Chris · 214
Thucydides · 88
Titus Livius · 52
Travis, Donald · 213
tribal chief · 46, 48, 49, 64
Tutankhamen · 59

U

uninhabitable cities · vi
Urban II (pope) · 61
urgency
out of control · v

V

value in society · 153
Verne, Jules
Mysterious Island · 181
Voltaire · 199

W

Walker, Katey · 214
Washington, George
as a terrorist · 4
as an insurgent · 4
Weaver, Henry Grady
The Mainspring of Human Progress · 32, 82, 152, 158
witch doctor · 46, 48, 49
World Trade Towers
time to build · 27
time to destroy · 27

Z

Zemansky, Mark W · iii

Made in the USA
Monee, IL
04 November 2026

67383830R00134